295

WILLOW TREE SHELL

GW

PYROTECHNICS

by
GEORGE W. WEINGART

Second Edition,
Revised and Enlarged

1947
CHEMICAL PUBLISHING CO., INC.
BROOKLYN   N. Y.

PRINTED IN U.S.A.

PREFACE

The favorable reception accorded by the public to the previous writings on the subject of fireworks-making by the author has induced him to prepare a revised and enlarged edition of the book which brings up to date the data on an industry which is becoming increasingly important.

There have been writings on the subject of pyrotechny as early as the year 1225. The first authentic manual is in German, written in 1432. Since 1893 some historical and specialized articles have appeared in connection with pyrotechny, but nothing of a general or comprehensive character in the way of a working manual has been published in English since Thomas Kentish's *Pyrotechnists Treasury* in 1878.

Meanwhile, great strides have been made, both in the materials used and in the methods employed in mass production of articles in general use for public celebrations.

Powdered aluminum has added many beautiful effects; picric acid is used extensively. Potassium perchlorate has greatly reduced the dangers of spontaneous combustion. Phosphorus, both yellow and amorphous, is used to a great extent. Machines to replace manual operations have been devised in many instances.

It is the object of this book to cover very thoroughly the developments in this interesting industry, as well as to present the latest methods and products of manufacture. It places in the hands of the beginner a working manual which will assist greatly in the production of every known piece of fireworks. Details are explained in such a way that the operator is never left in doubt as to how to proceed. The methods employed in large factories are also outlined. Even the professional pyrotechnist will find useful information in this book.

iii

The formulas given are all taken from those in actual use and will positively produce the effects for which they are indicated. Quite a few persons have written to ask whether the proportions, as indicated in formulas, are for pounds or other units. Since nearly all mixtures for fireworks call for dry materials it makes no difference what units is used provided the *same* unit is employed throughout. For instance, if a composition calls for 4, 2, 1, of the ingredients, 4 pounds, 2 pounds and 1 pound, or that many ounces or any other unit of weight, may be used. Where an exact amount of a liquid is indicated, the other ingredients also are specified in pounds, ounces, etc. The machines and tools shown are those in use, both by beginners and in the largest and most recently equipped factories.

The prices, which are indicated in the first part of the book for the various substances used in fireworks-making, have, in most cases, increased considerably, due to present abnormal conditions. Nevertheless they have not been changed in this book in the hope that they will soon return to their old level.

When using fireworks be sure to take the following precautions: Never hold any kind of fireworks in the hand after lighting. Make this a rule without exception because sometimes the most unexpected explosions occur and cause painful injury. In firing shells never look into the mortar after the shell has been inserted. A spark might have gotten inside and will cause a premature discharge. Also never allow the arm or hand to project over the muzzle of the mortar. Long sections of bare match should be used on shells so that the one lighting them has ample time to reach a safe distance in case of the bursting of the gun. There is always enough danger without taking unnecessary risks.

The beginner is advised to read carefully and memorize the instructions given under *Manipulations* before attempting to make any kind of fireworks. Actions, which seem trivial, are sometimes fraught with danger. The object of fireworks is to give pleasure, therefore, do not compound it with tragedy.

In conclusion I might add that, in this work, pyrotechny is treated rather as a craft than an art. For the artistic feature, imagination is the only requisite but to produce the results of imagination is tedious, patient and often hard work. Consequently, if we have the methods at hand it will be easier to fulfill our artistic ideas.

G. W. WEINGART

TABLE OF CONTENTS

INTRODUCTION

To THOSE contemplating the making of fireworks, either professionally or for pleasure, it is desirable to understand the principles which govern the operations of the various devices as well as the compositions of the chemicals used for their production.

It is not sufficient for the pyrotechnist to know what mixtures of various substances will produce the effects which he desires; he should also understand the reasons *why* these effects accrue.

For instance, he may know that a mixture of saltpeter, sulfur and charcoal will explode when a lighted match is brought in contact with it, but he should also know why it explodes.

Broadly speaking, practically all pyrotechnical compounds owe their action to chemical decomposition. This may occur under four different conditions: rotting, burning, explosion and detonation. The second and third of these are made use of most frequently by the pyrotechnist. The last, with a few exceptions, he tries to avoid and the first is of no value to him.

Rotting is a slow process, usually produced by fungi and bacteria aided by moisture and warmth. Burning proceeds much faster and one of the objects of the pyrotechnist is to control its speed. Explosion is due to a violent reaction of the chemical elements and it is usually brought about by the application of heat. The requisite heat may arise from fire, friction or spontaneous combustion.

Detonation is an instantaneous decomposition or reaction of the substances involved. This phenomenon is made use of in blasting with dynamite, etc., where the greatest possible energy is desired from the substances involved. It is brought about mostly by the use of fulminates, which themselves detonate when ignited or struck, communicating their action to the substance which it is desired to fire. Dynamite, gun cotton, etc., simply burn when lighted but when a small amount of fulminate is detonated in their midst violent action is produced.

Chemical reactions proceed only under certain conditions. For instance, if one part of oxygen and two parts of hydrogen gas are mixed in a container at ordinary temperature nothing occurs; but if a spark is produced in this mixture the reaction proceeds with a violent explosion and the production of water. This principle applies to practically all pyrotechnical mixtures.

The production of colored lights is based on:

1. Producing a mixture that will burn at a reasonable speed while generating intense heat.

2. Adding to it the salts of elements in the spectra of which lines of the desired colors predominate.

Heat generating compounds consist chiefly of:

1. Substances which yield oxygen freely when ignited in the presence of carbon, e.g.:

Potassium chlorate	Sodium nitrate
Potassium perchlorate	Barium nitrate
Potassium nitrate	Strontium nitrate
Sodium chlorate	

2. Carbon and carbonaceous sources:

Charcoal	Stearin
Shellac	Milk sugar
Fossil gums	Sugar
Resins	Flour
Asphalt	Petrolatum
Dextrin	

In addition to the above, there are some substances which, when added to colored fire compositions, increase the affinity of the constituents for one another or generate more heat, and, in that way, improve the colors:

Sulfur

Picric acid

In the case of blue and green fires it is almost essential, in order to get sufficient depth of color, to add an easily volatilized chloride such as:

Calomel

Sal Ammoniac *

The exact function of these two substances is not entirely clear but it appears that the best spectra are yielded by chlorides. However, most of the chlorides are deliquescent and therefore unsuitable for fireworks. By adding a substance that yields chlorine freely, at the moment of decomposition, the necessary conditions are produced for obtaining the best results.

The following substances are most generally used for producing pyrotechnical colors:

RED

Strontium nitrate

Strontium carbonate

Lithium salts (rarely used)

PINK

Calcium carbonate

Calcium sulfate

Calcium oxalate

GREEN

Barium nitrate

Barium chlorate

Boric acid and thallium salts (rarely used)

BLUE

Copper carbonate

Copper arsenite

Copper sulfate

Copper sulfide

Copper black oxide

Copper oxalate

Copper and ammonium sulfate

Copper and ammonium chloride

---

* The use of ammoniacal compounds with chlorate of potash requires the greatest caution and is not advised.

YELLOW

Sodium oxalate
Sodium bicarbonate
Sodium metantimonate
Sodium nitrate
Sodium perchlorate

ORANGE
Strontium and sodium salts

PURPLE
Strontium and copper compounds with calomel

The materials used for making *plain* or *bright* mixings are:

Saltpeter	Zinc powder
Lead nitrate	Antimony
Sulfur	Orpiment
Charcoal and lampblack	Realgar
Steel filings	Aluminum
Iron borings	Magnesium

Part I

INGREDIENTS

Saltpeter

(Niter—Potassium nitrate)

THIS most important ingredient of fireworks is produced in the United States. The American product is of the required high quality. The most suitable for pyrotechnical purposes, in general, is the *double refined powdered* saltpeter which can be obtained at 5 to 15¢ a pound in barrels of 350 pounds, the price varying according to market conditions.* For some large work, granulated saltpeter is used, since it is both slower-burning and cheaper. The most suitable is Dupont #2.

Specifications for saltpeter to be used in fireworks call for a salt that is clean, white and fine enough to pass through an 80 to 100 mesh sieve. It should contain less than 1% of sodium, calcium and magnesium salts combined.

Potassium Chlorate

The manufacture of potassium chlorate in commercial quantities made possible the production of beautiful pyrotechnical colors. The potassium chlorate prepared in this country is of excellent quality. The price varies considerably and it usually declines somewhat after July 4th. The price of the powdered form ranges from 9 to 16¢ a pound in kegs of 112 pounds. For pyrotechnical purposes, it should be white, odorless and contain not over ½ to 1% of sodium, calcium and bromine combined. It should pass through an 80 to 100 mesh sieve.*

---

* See preface.

* The danger of spontaneous combustion when using this salt in combination with sulfur is greatly reduced by the addition of a little sodium or potassium bicarbonate.

Potassium Perchlorate

This substance is a valuable addition to the art of pyrotechny. Although it contains even more oxygen than the chlorate, it is less liable to decomposition, due to the fact that it is a salt of perchloric acid which is a much more stable acid than the chloric acid from which the chlorate is derived. It can be substituted for the chlorate in most mixtures and can be safely used in combination with sulfur. However, stars and similar fireworks made with it are much more difficult to ignite and starting fire is frequently necessary. Addition of powdered charcoal sometimes is of assistance. The price is slightly higher than that of chlorate and its specifications are practically the same.

Sulfur

High-quality *flour of sulfur* is made in the United States from the product of Louisiana and Texas mines and is sold at 2 to 3¢ a pound in barrels of 250 pounds. Italian washed sulfur is recommended by some of the older English pyrotechnists for use with chlorate of potash but the writer tries to avoid the use of formulas containing the two substances and with the advent of potassium perchlorate there is no need for using them. Flour of sulfur is almost white. *Flowers of sulfur* and coarsely ground sulfur are also used. Coarsely ground sulfur burns somewhat more slowly than the other two varieties. In slow-burning torches, ground sulfur is frequently used as the large lumps seem to have the tendency to cause the *chimney,* which often forms on these lights, to break off. Specifications call for less than 0.1% of impurities and the finely ground sulfur should pass through a 120 mesh sieve.

Charcoal

Willow charcoal is the best for fireworks, although charcoal made from any soft wood is suitable. Pine charcoal is not very desirable. Excellent charcoal is made and powdered near Rochester, N. Y. It is available in finely powdered or granulated form or mixed, as desired, at about 1½¢ a pound in barrels or sacks. A brown tint usually indicates incomplete carbonization and such charcoal should be avoided. It should contain a minimum of grit. Shaking a sample in a bottle of water and decanting several times will disclose an excessive amount of sand, etc.

Lampblack

Germantown lampblack is very popular with fireworks makers, but, there are any number of good brands on the market. To make a good bright star, the lampblack should be free from oil or other impurities. It is sometimes necessary to bake it by packing it into a large crucible or iron pot and heating to redness, being careful to keep the pot covered to avoid burning. This will rid it of volatile impurities which impair brilliancy when burning. It can be bought in 1 pound packages, packed in barrels at 3¢ a pound.

Shellac and Other Gums

Shellac, a gum-like substance, is the secretion of an insect living on a large variety of trees in northern India. After going through various processes, it finally reaches this country in about a dozen different grades. The T N is a good grade for fireworks and the finely powdered shellac costs 18 to 25¢ a pound when bought in large quantities. It is practically impossible to powder this article manually, a ball mill being necessary. Great care must be exercised in purchasing it as it is frequently adulterated with sand, etc.* For the best work, shellac is almost indispensable.

For stock goods, tableau fires and torches, a number of substitute gums have been introduced such as Kauri, a fossil resin

---

* It may be tested by mixing it with water in a beaker. The particles of shellac will float, whereas, the sand, etc., will sink.

of light-yellow to dark-brown color, obtained from New Zealand. Red gum comes from the Kangaroo Islands 5 miles from Australia and costs about 5¢ a pound. K.D. dust is used for green fire. Asphaltum produces excellent colors when finely ground, but owing to its sulfur content, or perhaps because it is so easily decomposed, it is liable to spontaneous combustion when mixed with potassium chlorate. A mixture of these will explode violently when struck with a hammer on an anvil. With potassium perchlorate, however, it is safe. The Syrian asphalt is the best. The so-called green gum consists of powdered coconut shells and has no more value in pyrotechny than sawdust. Flour, milk sugar, dextrin, etc. are frequently used as sources of carbon.

Stearin
(*Stearic Acid*)

Stearin is another substance used as a source of carbon. In making blue fire, it has been found that stearin produces a better effect than any other hydrocarbon, especially with Paris green and other copper salts. It is usually obtained in cakes and is reduced to serviceable condition by setting a carpenter's plane upside down over a box and shoving the cakes against the blade so as to shave the stearin as finely as possible. Then, when it is mixed with the other ingredients it will pass through an ordinary sieve.

Strontium Nitrate

This substance is made principally in England. Its chief source is celestite. Strontium nitrate is put up in 110 pound kegs and casks of 600 pounds in a sufficiently pure condition for use, costing from 6 to 9¢ a pound. It is probably the most useful color-producing chemical used in fireworks as the deep-red light which it gives is the most marked effect the pyrotechnist has achieved. A number of methods has been devised to overcome its deliquescence, one of which is to melt some shellac in an iron pot over a fire and stir in the nitrate, pulverizing it when cooled. Another plan is to use strontium carbonate, but at the cost of considerable depth of color. Strontium nitrate is used as a somewhat coarser powder than the potassium salts but should be clean and white and contain not over 0.2% of moisture and 0.25% of sodium salts.

Strontium Carbonate

In damp climates, there is no alternative but to use this strontium compound. A piece of lance work made with nitrate of strontium, if exposed for 1 hour to a very damp atmosphere, will hardly burn. Precipitated strontium carbonate is the only kind which should be used and may be purchased for about 16¢ a pound or can be easily made by adding carbonate of ammonia to a solution of strontium nitrate, then thoroughly washing and drying the precipitate. If sodium carbonate is used as a precipitant it is almost impossible to remove every trace of sodium from the strontium carbonate and the latter gives an orange tint to the red light.

Barium Nitrate

Like strontium, this chemical comes to us mainly from England and Germany in similar packings and costs in normal times from 5 to 7¢ a pound in the United States. As a color producer it is very inferior to strontium, but it does not attract moisture. If used without calomel its color is very pale. Specifications for fireworks are practically the same as for strontium although it comes in a much finer powder.

Barium Chlorate

This is a better salt for making green fire and gives a very beautiful emerald color. However, its high cost, i.e., about 30¢ a pound, prevents its use except in exhibition work. It is also rather sensitive and easily decomposed and great care must be exercised in handling it. Some formulas for green fire use

Pyrotechnics

boracic acid, thallium salts, etc., but if these are used at all it is to a very limited extent. All barium salts are very poisonous.

Sodium Oxalate

It is a strange fact that although yellow is the most common color of fires in general its practical production in pyrotechny is accompanied by some difficulty. There is practically only one water-insoluble salt of sodium, all the others being more or less hygroscopic. The nitrate and bicarbonate give fine deep-yellow lights, but the least dampness will render them incombustible and even the oxalate will attract moisture in damp weather. Perhaps, it would be more correct to say that the very weak oxalic acid is easily replaced in the presence of moisture by the stronger nitric or chloric acid of the other ingredients. This converts the sodium oxalate into a hygroscopic salt which causes other salts in the composition to decompose, rendering the mixture useless. The exception in this case is sodium metantimonate, but as this salt costs $4.00 a pound and gives a pale color, it is not used to any great extent.

Sodium oxalate costs about 20¢ a pound and can easily be made by adding sodium bicarbonate to a hot concentrated solution of oxalic acid. A copious precipitate is formed which, however, should not be washed but dried on the filter. An excess of oxalic acid should be used for the precipitation.

Copper Arsenite
(*Paris Green*)

This compound is made in the United States and can be bought for from 10 to 15¢ a pound from dealers of paint supplies. It can easily be made by adding a solution of blue stone (copper sulfate) to a solution of arsenous acid in water; the resulting bulky precipitate should be washed and dried. It is used in making blue fire. Copper arsenite, used for making green paint, is also satisfactory for fireworks.

Copper Carbonate

This substance is also used for making blue fire, but better results are obtained with less trouble by the use of other copper compounds. The native carbonate is almost useless for the manufacture of fireworks. The precipitated carbonate, which is more suitable for this purpose, is easily obtained from dealers in pyrotechnical materials. It can be made by adding ammonium carbonate to a solution of blue stone. Chertiers copper is made by carefully adding aqua ammonia to a solution of blue stone, evaporating and crystallizing. The author has obtained the best results with the double salt of copper and ammonium chloride, which may be had at a reasonable price and does not require the addition of calomel.

Copper Sulfate
(*Blue Stone*)

For most purposes, where a good blue was required for exhibition purposes the older pyrotechnists used this salt, but owing to its tendency to oxidize with the liberation of sulfuric acid, great care must be taken in making mixtures containing chlorate of potash. Separate sieves should be used for mixtures of these substances, which should not be used in any other work. It usually costs 10 to 15¢ a pound. Mixtures containing it must not be stored but should be used promptly after making. Mix and sift all the ingredients, except the chlorate of potash, first; then add the potash and mix again. Otherwise, heat is generated during the mixing. It may be used safely with potassium perchlorate, but stars made with this salt are difficult to ignite.

Black Copper Sulfide

This compound, when used in conjunction with calomel, is valuable in the production of blue and purple fires. It is important to note in this connection that only the product made by fusion is of value in pyrotechny, the precipitated black sulfide being useless. It is sometimes difficult to obtain, but it is quite easy to make. The following method is suggested:

Procure thin sheets of scrap copper and cut them into pieces about 1 inch square. Pack a large clay crucible with alternate layers of copper scraps and powdered sulfur, to within an inch of the top. Cover and place in a bright-red fire for about 1 hour. When cooled the contents may be shaken out and ground or pulverized for use. They should be screened through a 60 to 80 mesh sieve. Exact proportions of sulfur and copper are not essential as long as an excess of sulfur is present. This excess will burn off in fusing.

Black Copper Oxide

This is used similarly to the above and is easier to obtain. As in the case of the sulfide, only the fused form may be used. The brownish-black, light, precipitated oxide is useless.

Antimony

Metallic or regulus antimony, finely powdered in an iron mortar, is used in making white fire. It may be had from machinery dealers at 6 to 7¢ a pound.

Magnesium

After being unused by the pyrotechnist for 40 years, this metal has again entered the picture. It is much more inflammable than aluminum. A thin strip of the metal when lighted with a match burns with an intense white light. Its principal use is in flash light bulbs for photography and in incendiary bombs for warfare. When bombs made of it burst, each fragment takes fire and burns with great heat.

Its price has been reduced below that of aluminum, i.e., 26½¢ per pound in large quantities. It is also lighter in weight than aluminum.

Black Antimony Sulfide

This may be obtained from most drug dealers and costs about 6 to 8¢ a pound. It is sometimes very impure and entirely useless. As low as 70% purity is serviceable for pyrotechnical purposes and it is used for making white fire, maroons, smoke effects, etc. Red and orange sulfurets are also used. The compounds are poisonous.

Red Arsenic
(*Realgar, Orpiment*)

These minerals come mostly from Hungary in kegs of several hundred pounds, ranging from 6 to 9¢ a pound for the powdered product. Although these minerals are quite poisonous, the miners become immune to them and even eat them to ward off disease. They are very useful in making white stars, especially as they take fire far more easily than those made from antimony. Arsenic compounds are also used in making yellow smoke in day fireworks.

Aluminum

When, about 60 years ago, it was found that a star of unusual brilliance could be produced by the use of magnesium, that metal suddenly came into considerable demand, in spite of its cost of $75.00 a pound. Later, however, it was found that aluminum was in every way better and could be obtained for about 60¢ a pound in various degrees of fineness from an impalpable powder to the *flitter* which is quite coarse. It may be obtained from most paint dealers in cans or in 1 pound paper packages. It should be marked *C.P.* as there is some on the market which is adulterated with finely powdered mica which renders it useless for fireworks. It should be at least 95% pure and may contain 2% of fatty material. The *pyro* is a dark powder. It is used in flash crackers, and the utmost care must be exercised in handling it; compositions containing chlorate of potash and aluminum are easily fired with disastrous results. Only the perchlorate is recommended.

Added to stars and torches, aluminum greatly increases their brilliancy and beauty. Exquisite waterfall effects are produced with it, as well as comets, tailed stars and intensely bright flares. Large quantities of the finely divided aluminum (pyro aluminum) are used in flash crackers and maroon shells. Besides increasing the report, it gives a startlingly bright flash to the explosion. Being unaffected by water, it is much safer than magnesium, but care must be used in handling it because, as previously mentioned, all finely divided metals are liable to explode when in contact with oxygen-producing chemicals. Rubbing a small quantity of petrolatum into it seems to reduce the danger of accident.

Calomel

This is used to deepen the color of fires. As previously stated, it has been found that the chlorides of the metals give the best spectra, but as chlorides are usually not practical for manufacturing fireworks, the addition of an easily decomposed chloride produces the necessary conditions at the moment of combustion. The reason that copper compounds by themselves give off a green light yet in combination with calomel, etc., burn blue is too complicated to be described here. Calomel is made in this country and sold in normal times at about 65¢ a pound, but due to the scarcity of mercury and the great demand for it in the manufacture of detonating caps the price has recently advanced to several dollars per pound.

Ammonium Chloride
(*Sal Ammoniac*)

This is sometimes used as a substitute for calomel, but its affinity for moisture seriously interferes with its general use. The crystallized salt is practically useless. It should never be used in combination with chlorate of potash and in any case should be used only in articles for immediate consumption rather than in those to be stored.

Dextrin

In all the old works on pyrotechny either a solution of shellac in alcohol or gum arabic in water was suggested to bind compositions for making stars and the like. At present the necessary amount of dextrin is added at once to the mixture and then nothing but water is required to form it into the desired objects. Dextrin also improves the color of some fires and it may be advantageously used in place of glue for light work. Potato dextrin usually comes in sacks of about 200 pounds and costs from 2½ to 5¢ a pound. When used for gumming rocket sticks, tabs, etc., it is simply mixed with water to the desired consistency. The light-brown #152 is most suitable for pyrotechny.

Glue

Several forms of glue are used in making fireworks. A good grade of carriage glue is best for attaching lances to framework. For attaching mine bottoms, etc., to cases, cheap carpenter's glue will suffice. For placing shell fuses and securing the ends of cannon crackers a good liquid glue is most convenient.

Gum Arabic

In powdered form, this is used in some star compositions, especially for making Japanese stars. It is also used in *Son of a Gun* composition, in *exhibition match* sparklers, etc.

Steel Filings
(*Cast Iron Borings, etc.*)

A beautiful scintillating effect is produced by using steel filings in various ways. The Japanese make a little tube of twisted paper, at one end of which is a composition which, when lighted, produces a glowing bead of molten flux. The balance of the tube contains steel filings which, when reached by the fused bead, burst into beautiful feather-like flashes. In other countries, steel filings are added to gerbes, fountains and driving cases with resultant brilliancy. A beautiful *Niagara Falls* effect is produced by charging from 50 to 200 cases (2 inches in diameter and 12 inches long) with a composition containing cast iron borings. These are fastened to scantlings at intervals of 15 inches, each scantling holding 16 gerbes. These are matched and hoisted to a wire cable about 50 feet above ground. When they are burned the effect is most realistic as the arc of the suspended wire has just the right perspective. The roar of the burning gerbes is also characteristic of Niagara as the fire from the iron borings drops to the ground.

The best type of steel filings for gerbes is known as needle steel. This resembles broken sewing machine needles but is really a by-product of some steel-planing operation. The steel filings from hand-saw filing shops are good, provided they are clean and are not particles thrown off by emery wheels.

When steel filings are added to gerbe compositions the saltpeter quickly attacks them, frequently causing the gerbes to become quite hot. The steel is rusted and this action practically destroys its usefulness. To prevent this, the steel must be coated in some way so that the saltpeter cannot reach it. This may be accomplished as follows: In an agate-ware saucepan place a piece of paraffin, melt it carefully and heat it as hot as possible without permitting it to smoke. To this add as much of clean steel filings as the paraffin will thoroughly coat. There should be no surplus paraffin, but just enough to completely cover each filing. Shake the pan and stir frequently while cooling to prevent the filings from caking. Steel filings are also used for stars, rockets and shells.

Clay

This is used for closing the ends of most cases as well as choking them when they are not crimped. Almost any kind of clay will do. It must be thoroughly dried, pulverized and sifted. Before use, it may be slightly dampened although this is usually not essential. For convenience, where a large quantity is required, powdered fireclay in barrels may be used. This saves the rather tedious job of drying and powdering.

Gun Powder

This is used in pyrotechny in various ways:
1. For driving shells
2. For propelling the stars in Roman candles
3. For producing the report or *bounce* at the end of gerbes
4. For making match
5. For bursting charge in shells, etc.

In all cases, with one exception mentioned later, the rifle or Blasting A, in gradings from FFFF for shells to FFFFFF for Roman candles, is used. This powder is made with saltpeter. B Blasting powder is made from sodium nitrate and is considerably cheaper when large quantities are used. However, it is also hygroscopic and therefore not recommended for any purpose except driving shells. For this purpose F powder, having grains about the size of cracked corn, is suitable. Because it burns much more slowly than the finer-grained powder, a larger charge may be used with less danger of bursting the shell in the mortar; besides, the shell can be driven to a greater height.

Both kinds of powder come in 25 pound iron kegs.

Meal Powder

This material is widely used in display work for gerbes, etc., as well as in shells and rockets as a blowing charge. It is also used for making match and priming, as it moistens more quickly than grain powder. It is generally supplied in 25 pound wooden kegs but is sometimes difficult to obtain outside of large cities. In that case a fairly good meal powder can be prepared in the following manner:

Mount a 50 gallon wood barrel on two uprights so that it can revolve freely on centers fastened to the heads. To one center attach a crank. Cut a hole in the side for putting in and removing the necessary ingredients, this hole should be closed by a suitable plug. Place in the barrel from 300 to 500 lead balls about 1 inch in diameter. Then add a thoroughly mixed composition as follows:

Saltpeter, double refined	15 lb.
Willow charcoal	3 lb.
Sulfur flour	2 lb.

The barrel is now revolved for about 500 turns. The longer it is turned the stronger the powder will be. Great care must be exercised to see that no foreign matter, such as nails, gravel, etc., finds its way into the barrel as this might result in an explosion. In fact, it is desirable to turn the barrel by a small motor or better still, by a water wheel. The entire procedure should be conducted at a distance, by remote control, as is done in regular powder mills.

Other Ingredients

Picric Acid

Picric acid is another valuable ingredient of fireworks. When added in small quantities to colors it deepens them and increases their brilliancy without making them burn much faster. Beautiful colors, which are almost free from smoke, can be produced with it. It is also used in *black snakes*. But it must always be kept in mind that picric acid (tri-nitro-phenol) is related to TNT, the tremendous explosive force of which is only too well known. For this reason, it cannot be used in shells, as stars made with it will detonate when confined instead of burning. Another article for which large quantities of picric acid were used until several years ago, when a fatal accident occurred in a factory engaged almost exclusively in its manufacture, is the amusing *whistling* fireworks. Picrate of potash has the peculiar property of emitting a shrill whistling sound when rammed tightly and burned in a small tube. If made in small quantities and carefully handled it seems to be reasonably safe, but the result of a barrel of it, accidentally detonated, can readily be imagined. This product was used in the *screaming bombs* in the beginning of World War II for the purpose of lowering the morale of non-combatants and weakening their nerves.

Yellow Phosphorus

Still another substance producing a most beautiful effect when fired from specially prepared rockets, as will be explained later in detail, is yellow phosphorus. It is with this that the liquid fire rockets are made and a more beautiful display does not exist. The rockets have a deep-yellow flame and as it falls through the air, it breaks into myriads of incandescent particles with a heavy backdrop of white smoke. Obviously, the greatest care must be exercised, as phosphorus burns, even when very small, are most painful. On the other hand, its use results in no *surprise features* and, if its nature is understood, it is less dangerous than many other things. A considerable amount of phosphorus is used in the manufacture of the *Son of a Gun, spit devil* or *devil on the walk,* etc. Poisoning and some deaths of children, who mistook the tablet-like contrivances for candy, have caused their restriction in some states.

Amorphous Phosphorus and Silver Fulminate

Amorphous phosphorus is the base of most of the toy torpedoes in use. Fulminate of silver was used almost exclusively for this purpose 60 years ago, but now only a small amount is used owing to its very sensitive nature. However, its method of use and preparation will be given later as a matter of record.

Zinc Powder

Zinc powder is used to some extent for making what is known as *Electric Spreader* stars. These produce an original effect, breaking up while burning into many small bluish-green particles. These are propelled with considerable force, giving the appearance of electrical discharges, hence the name. On account of the explosive nature of zinc dust, this star must be made with caution and reserve until it is well understood.

Substitution

It should be kept in mind that a good pyrotechnist must not be confined to a single formula or chemical to produce a given effect. Of course, there is always a *best* mixture or chemical for making a star, colored light, etc., but it sometimes happens that the particular substance necessary is not available at the moment. In that case, the next best must be used and it is up to the pyrotechnist to discover this.

In the introduction to this book is given a list of the various substances available for supplying oxygen, sources of carbon and the various metallic salts for producing the different colors, etc. When a particular substance cannot be obtained, often one of a similar nature may be substituted with some alterations of the formula. This is sometimes made necessary by the location of the plant. Salts, particularly of sodium, which may be used to advantage in a dry western atmosphere are useless east of the Mississippi River. Stars, especially those containing lamp-black, require a week of good weather to become thoroughly dry; they may appear to be dry to the touch and still may contain sufficient moisture to prevent satisfactory performance. Stars or similar preparations should not be made in very cold or rainy weather.

Part II

MANIPULATION

THE handling of explosives, naturally, is never entirely safe. No more so are electricity, gasoline and many other things in daily use in this modern age. However, many persons have devoted long lives to the manufacture of fireworks without having an accident. Nevertheless, even with the greatest care, accidents will occur, to both those engaged in making fireworks and to those burning them. An attempt will be made here to point out the usual sources of accidents, although, obviously, it is impossible to foresee every instance in which some carelessness or unknown factor may bring on disaster.

First, separate places are required for the making of so-called *plain mixings* containing sulfur and those mixings containing chlorate of potash. All utensils employed for the one type of mixture must be kept completely separate from those used for the other type of mixture. Persons employed in the *plain* sections of the factory must not go into the rooms of those in the *colored* section.

Second, keep in mind that very slight friction will sometimes start the burning of mixtures of finely divided chemicals. Star composition has been known to explode while being sifted by the scratching of the brass wire sieve bottom with the finger nail. Rockets have taken fire from the brass solid rammer striking the top of the spindle.

Third, finely divided metals, as previously mentioned, when in contact with chlorate of potash sometimes take fire suddenly during the mixing process. Fortunately this is seldom the case; nevertheless one must not lose sight of this possibility. Steel

17

filings and iron borings frequently evolve heat when mixed
with saltpeter or rammed into gerbes, thus creating the possibility of fire. In a similar way, shell fuses have exploded when
rammed too hard. The prevention of these occurrences is explained in the section entitled "Steel Filings."

Employees in the ramming rooms of factories should be required to wear rubber overshoes while at work and should not
carry matches. This can be partly controlled by requiring the
employees to change their clothes in the factory before going to
work and having them wear garments without pockets, but they
will sometimes slip out for a smoke during rest hours and have
matches with them. Close supervision is a necessity.

Small buildings about 12 feet square and not less than 50
feet apart should be used to house those engaged in mixing and
ramming operations as well as those making stars and, if possible, there ought to be only one person to a room. Doors should
be placed at both ends of work rooms and should open to the
outside, with no fastenings on the inside. They should be held
closed by spring hinges. Fire buckets, inspected daily, should be
in each building, supplemented by fire hose conveniently placed
for emergency.

The most successful method of reducing serious accidents is
to keep the least possible amount of composition on hand in the
work rooms. Remove to storage or finishing rooms all rammed
articles as quickly as they accumulate.

Striking fire is another source of danger against which the
pyrotechnist must be on constant guard. It is usually caused by
steel tools being struck together in the presence of flammable
compounds. Sometimes brass and wooden parts produce enough
friction to cause fire.

Spontaneous Combustion

Potassium chlorate, one of the most useful of fireworks
chemicals, causes spontaneous explosion. This is due to the
fact that its acid component, chloric acid, is an unstable
acid and is easily decomposed. Consequently, a slight rise in

temperature is sometimes sufficient to bring about an explosion. The tendency of potassium chlorate to explode is very strong in the presence of sulfur, sulfides and sulfates which sometimes release minute quantities of sulfuric acid. For this reason, compositions containing both chlorates and sulfates should be strictly avoided. All the chlorates act in a similar way.

Finely divided metals combine with oxygen easily and sometimes the reaction is so violent that they take fire at ordinary temperatures. Aluminum and zinc are examples. Aluminum and chlorate of potash may at times be subjected to violent friction and even struck with a hammer on an anvil without effect; yet at other times a mixture of these two substances has exploded without any apparent reason, causing fatal accidents. Since these substances are so easily ignited, there is a possibility that a minute electric spark may cause an explosion during sifting. Until the reason for their hazardous nature is better known, mixtures of these substances should be avoided.

Long experience has shown that the following mixtures behave as described here and should be handled accordingly.

Saltpeter, sulfur, charcoal and lampblack are the safest materials used in fireworks. Accidents with these occur only when a spark has been struck in some manner and brought in contact with the mixture.

Barium nitrate is not reported to have caused any accident with sulfur, saltpeter and aluminum.

Barium and strontium nitrates and aluminum and potassium perchlorates are among the safe mixtures even when used in combination with sulfur and gums. Barium and strontium nitrates, in combination with potassium chlorate and shellac or other gums, form a sensitive mixture and this condition is increased when powdered charcoal is added. Therefore, all unnecessary friction should be avoided during handling.

Barium chlorate yields oxygen quite readily so that it must be handled with great care in mixtures containing shellac and other hydrocarbons.

Aluminum powder in mixtures with potassium chlorate,

barium nitrate and shellac or other carbon sources is classed as *hazardous*.

Mixtures of potassium chlorate with sulfur, sulfates or sulfides, as previously mentioned, are to be avoided at all times (see note under Potassium Chlorate in Part I). Records show that 90% of the accidents in factories have been traced to this source. Chlorates should never be mixed with ammonium salts, as this combination often results in spontaneous combustion.

Potassium permanganate and bichromate mixtures should be handled with great care, especially in combination with finely divided metals.

Mixtures of picric acid and chlorates are too sensitive for ordinary use.

Compositions containing potassium chlorate and phosphorus must never be mixed except under water. Yellow phosphorus must not be removed from water for more than a few moments at a time and then handled so as to avoid friction or proximity to combustible materials (wood, paper, etc.).

Those using red phosphorus should never be permitted to enter the factory where other fireworks are being made, as even a few grains of this substance in the clothing are sufficient to cause an accident when coming in contact with other substances.

Fulminates of mercury and silver should only be handled by those familiar with their properties.

Care is urged in the use of the foregoing combinations, not so much when small quantities are used but more specifically when amounts of 10 to 100 pounds are used. They should never be handled thoughtlessly and treated as just so much sand or cement.

When experimenting with new substances use the smallest possible amounts of the component chemicals until the safety of the entire mixture is assured. Before using considerable quantities of a new composition, it should be subjected to exhaustive tests such as friction, percussion, detonation and the addition of moisture with subsequent drying. Also, the temperature of the flash point should be ascertained with suitable apparatus.

In testing colors, the pyrotechnist should not look directly at the mixture when it starts to burn but should have his back turned to it while someone else lights it. He should then turn quickly around for a moment, look at the light and then turn away again. Looking at the color immediately as it is being produced seems to affect the optic nerve temporarily so that an accurate appraisal cannot be made. It is also advisable to view the color from a distance of about 100 feet to judge it correctly.

Mixing

The first operation in the manufacture of fireworks, and undoubtedly the most important one, is mixing. Chemicals are available in the required grade of purity and in powdered form, so that long discussions on purifying, powdering, etc., are unnecessary. All chemicals should, of course, be obtained in the best quality available at a reasonable price and if possible finely powdered. Chemically pure drugs are not necessary.

For mixing on a small scale, round brass wire sieves are the best. For lances and the most particular work #22 to #26 meshes may be used. For plain mixtures #16 to #18 meshes are suitable. When 15 to 25 pounds or more are to be mixed, ordinary wooden wash-tubs are most convenient and the sieves should just fit inside the upper edge. For mixtures from 100 pounds up troughs are often used. For these the sieves are made square and fit inside of the troughs, resting on lugs. Mixing machines are sometimes used for *plain* mixtures containing no chlorate of potash.

For the plain mixtures the coal is weighed first and put on the bottom of the tub; the sieve is then put in place and the sulfur, saltpeter, etc., pushed through it. When everything is sifted, mix well in every direction. The bare arms and hands may be used, but rubber gloves are very desirable for this purpose. Now place the sieve on another tub of the same size and sift from the first tub into the second one, a scoopful at a time. When all of the mixture has passed through, mix again as before and sift once more into the first tub, giving a last thorough mixing. For ordinary compositions this is sufficient, but some mixtures are passed through the sieve four or five times.

In colored mixings more care must be exercised and each ingredient sifted separately the first time, except the shellac, coal, etc. These may be put right on the bottom of the tub. Never throw the chlorate onto the sieve with dextrin or other hydrocarbons. Sift the chlorate of potash first and add the other salts one by one. Great care should be taken never to let the finger nails strike the sieve while sifting, as it is very easy to *strike fire* in this way, with disastrous effects. Sharp star compositions in a loose state are almost as explosive as meal powder. Special mixtures will be described when we come to the compositions requiring them.

Case Rolling

Case rolling is the next important operation in the manufacture of fireworks and requires the most mechanical skill judging from the time required to learn it and the comparatively small number of really good case rollers to be found in factories.

All kinds of fireworks, except tableau fires, require a case of some kind. A good case must be tightly rolled and almost as hard as metal. The best arrangement for case rolling is a large sloping table made of tongue and grooved flooring, tightly joined and firmly nailed to sills of 2 inch stock, tapering from a height of 2 inches in front to 6 or 7 inches in the back so as to give a gentle rise from front to rear. According to the work to be done the rolling board may be made from 2 to 4 feet wide (see Fig. 1). A marble slab is very desirable for rolling rocket cases.

Most cases were formerly rolled from feathered-edge strawboard, the best being made in Elbridge, N. Y. It comes in

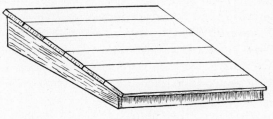

FIGURE 1

Rolling Board

sheets of 26 x 38 inches with from 40 to 150 sheets to a 50 pound bundle depending upon thickness. However, in recent years this has almost entirely given place to *waste paper board* made in the same sizes and weights. For rocket cases, 2 or 3 turns of hardware paper are used first, backed up with 5 or 6 turns of strawboard or other paper. The hardware paper, being waterproof, does not swell and contract in rolling whereas the strawboard, being absorbent, swells considerably; therefore when it is rolled on the outside of the case it contracts in drying, making a very firm case. Heavy Manila and so-called *cotton sampling* paper also make good rocket cases if carefully rolled, but as these shrink considerably in drying, the ramming tools are liable to stick unless they are specially adapted to this kind of paper or the formers are made correspondingly larger. The recently produced Kraft paper makes an excellent case, but as with the last two papers mentioned, very little and heavy paste must be used and the cases must be rolled as quickly as possible before the paper has time to soften. Straw and pulp boards cost about $25.00 a ton in normal times.

Lance Cases

The lightest cases used in the manufacture of fireworks are lance cases. Some pyrotechnists use poster paper of different colors for these, the colors indicating the color of the lances when burning. Other manufacturers use linen paper. The colored paper has the advantage of making lances easily distinguishable in case the boxes are mixed up. On the other hand, it requires keeping a larger stock of empty cases continually on hand. This is sometimes inconvenient. Linen paper is much stronger and only one kind is required, the lances of different colors being boxed individually with the colors marked on the outside.

Lances are made from $\frac{1}{4}$ to $\frac{3}{8}$ inch in diameter and from $2\frac{1}{2}$ to 4 inches long. Generally speaking, the greater the diameter the shorter the length may be. Generally, a lance case of $\frac{1}{4}$ inch in diameter is used. It is made of ribbed linen paper of 17 by 22 inches, about 16 pounds to the ream, cut in four, the smallest way (or across the ribs) and cut in 6 the long way (or with the ribs). This makes twenty-four cuts from each sheet, $3\frac{5}{8}$ by $4\frac{1}{2}$ inches. Now procure a brass or copper tube, with an outside diameter of $\frac{1}{4}$ inch, and some good paste. Take a bundle of approximately 1 or 2 dozen sheets and lay them squarely before you on the rolling board so that the $3\frac{5}{8}$ inch side is on top and bottom. Holding them down tightly with the left hand, rub them gently toward you with the thumb-nail of the right hand so that each one will slide about $\frac{1}{4}$ inch below and to the left of the one under it. Apply paste to these two edges and lay the tube now on the top sheet $\frac{3}{4}$ inch from the bottom and $\frac{1}{8}$ inch from the right, or pasted, edge. With the ends of the fingers of the right hand bend the lower edge around the tube, tucking it closely and roll up to the pasted end. Then with a turn of the fingers twist the bottom in. The bottoms should not be made too solid and if a little hole is left it will be easier to stick them on the pins. Sometimes when they are to be used for an exhibition made on the grounds and will not be handled much, the lance cases are made without any bottoms. They may now be thrown lightly in a basket or sieve to dry. These operations, although very simple, are quite difficult to describe and a few moments of practical demonstration will be much more beneficial than several pages of description.

Pinwheel Cases and Match Pipes

Pinwheel cases and match pipes are rolled, in general, in the same way as lances except that no bottoms are made to them and brass or steel rods are used instead of tubes for rolling them. The most convenient size for match pipes is 1 yard in length and $\frac{1}{4}$ inch in inside diameter. Use a good grade of Manila or Kraft paper 24 by 36 inches, 20 pounds to the ream. The quire is cut in the length, into strips 4 to 5 inches wide. A steel rod $\frac{1}{4}$ inch in diameter and 4 feet long is the best for rolling them. Pinwheel pipes are usually made 12 inches long and $\frac{3}{16}$ inch in diameter. Sometimes, one end is made slightly funnel shaped by pasting a strip of paper 6 inches long and 2 inches wide at one end tapering to $\frac{1}{2}$ inch at the other, rolled around one end of the rod, beginning with the 2 inch end and rolling so that the $\frac{1}{2}$ inch end terminates near the extremity of the rod. In use these cases are pinched, shut at the small end and bunched up with the funnel end up, making them somewhat easier to load. Rolling match pipes properly is one of the most difficult operations to master. It is therefore advisable to begin with shorter pipes until skill is acquired.

Roman Candle Cases

These are also difficult to roll and it is essential to have a feather-edged board for this work. The sheets for 1 to 4 ball candles are pasted entirely over with rather thin paste. A dozen or more are put down on the board at a time. Each sheet is placed half an inch nearer to the operator than the one under it. Workers roll 1 ball candles, 3 at a time, by pasting 3 rows of sheets on the board side by side. From 6 ball candles up, only about 4 inches on each end of the sheet should be pasted on both sides. Lay a sheet on the rolling board, well up near the top, with the feather edge nearest the operator. Then, with a 4 inch, flat paint brush apply thin paste quite heavily on about 4 inches of the top of the sheet and about the same amount on the bottom. Now, place another sheet on top of this but about 1 inch lower down, so that an inch of the first one extends beyond the next on top of it. Paste as before and repeat the operation until a dozen or more sheets are in the pile. Now, reverse the entire lot at once so that the former bottom sheet will be on top. Paste over the bottom and top edges of the pile now exposed and rub off surplus paste. Now you are ready to begin rolling (Fig. 2).

Lay the rod across the pile about 3 inches from the bottom. Lift the bottom edge of the first sheet, lay it over the rod, draw the rod (with paper over it) back until the edge of the strip is on top of the rod and slide the fingers along the rod and the edge of the sheet until the sheet sticks firmly to the rod for its entire length. Now, roll firmly along until the whole sheet is rolled up, with one hand following the other. Take care that the case does not run to one side or the other. By a quick twist of the rod in the opposite direction from that in which the case was rolled, the case can now be removed from the rod and placed

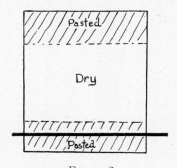

FIGURE 2

Sheets for Rolling Roman Candle Cases

on the rack for drying. Sometimes with beginners it happens that the case becomes so firmly attached to the rod that it cannot be removed. This results if the rod is too dry when the first turn is taken. Keep plenty of paste on the rod and this will not happen.

Of late, the diameters and lengths of Roman candles and rockets have been changed and reduced so often that no standard of sizes can be given, but the following will be found to be useful for average work and may be used accordingly. Special sizes may be easily adapted to the required circumstances. When cutting paper for candles and other cases as well, care should be taken to cut so that the case always rolls with the grain of the paper and with the feather edge at the top of the sheet.

No. of Balls	Length, Inches	Bore, Inch	Size of Sheet, Inches	No. of Strawboard or Waste Paper Board
1	4½	5⁄16	4½ x 6	140
2	5½	5⁄16	5½ x 7	140
3	6½	5⁄16	6½ x 8	140
4	8½	5⁄16	8½ x 10	140
6	12	3⁄8	12 x 13	140
8	15	3⁄8	15 x 16	140
10	17	7⁄16	17 x 20	140
12	19	7⁄16	19 x 20	140
15	22	½	22 x 26	120
20	26	½	26 x 26	120
25	32	½	32 x 26	100
30	36	½	36 x 26	100

Cases for Rockets, Gerbes, etc.

Cases for rockets, gerbes, fountains, tourbillions, saxons, etc., and the small paper guns used for mines, floral shells, etc., require considerable skill and strength for rolling, especially if they are of the larger sizes. After seeing a great many case rollers at work and employing, at different times, their various methods, the author has come to the conclusion that the following is not only the easiest, but the most efficient, method.

Take a small scrubbing brush of good quality with long, stiff hairs. Make the paste somewhat thicker than that used for candles. Lay a single sheet of paper on the rolling board (in the case of large rockets, include the sheet of hardware paper on top of it). Now, with the scrubbing brush, rub some paste (not as much as for candles) evenly over the sheet and immediately roll up as tightly as possible, all but the last 2 inches where more than 1 sheet is used. Now, paste the second sheet over as you did the first one and place the partly rolled case on top of it about 2 or 3 inches from the end nearest to you, seeing that the edges of both are even. Raise the end of the second sheet projecting behind the already partly rolled case and bend it around so that it will lay between the part of the inner sheet left unrolled and continue rolling forward, pressing the case firmly to the rolling board or marble slab until the case is completed. This leaves a case that is already half dry and, when completely so, should be firm enough so that it cannot be bent in on the ends with the fingers. The advantage of this method of rolling heavy cases is that there is no time for the paper, especially the straw or pulp board, to become softened and swoolen as is the case when a number of sheets are pasted down at once. A tighter, cleaner case, which is more easily and quickly dried, results. If too much paste is used, the water from the paste evaporates on drying, leaving the case spongy.

The sizes of rockets vary as much as those of candles. Consequently, the following list can only be used approximately.

Rockets

Size	Length, Inches	Bore, Inches	Length Sheet Strawboard or Waste Board, Inches	No. Strawboard or Waste Paper Board	Hardware Paper, Inches
1 oz.	$3\frac{1}{2}$	$\frac{3}{8}$	10	140	0
2 "	4	$\frac{3}{8}$	13	140	0
3 "	$4\frac{1}{2}$	$\frac{3}{8}$	$17\frac{1}{2}$	140	0
4 "	5	$\frac{1}{2}$	20	120	0
6 "	6	$\frac{9}{16}$	13	120	12
8 "	7	$\frac{5}{8}$	18	120	12
1 lb.	8	$\frac{3}{4}$	20	120	17
2 "	9	$\frac{7}{8}$	26	140	25
3 "	10	1	26	120	25
4 "	11	$1\frac{1}{4}$	26	120	50*
6 "	13	$1\frac{1}{2}$	52	120	50*

For the so-called 8 pound rockets the same cases can be used as for the 6 pound rockets.

* These can be conveniently used in two lengths.

MINES

No.	Height, Inches	Diameter, Inches	No. Strawboard	No. of Pieces
1	4	1½	140	1
2	4¾	1¹¹⁄₁₆	120	1
3	5½	2¹⁄₁₆	100	1
4	7	2¼	100	2
5	8½	2⅜	100	3
6	10	2⅞	100	4

FLORAL SHELL GUNS

No.	Height, Inches	Diameter, Inches	No. Strawboard	No. of Pieces	Length, Inches
1	9	2⁵⁄₁₆	100	3	26
2	11	2⁷⁄₁₆	100	4	26
3	13	3³⁄₁₆	100	4	26
4	15	3½	100	5	26

GERBES

Length, Inches	Diameter, Inches
9	¾
11	1
13	1¼
15	1½

NIAGARA FALLS

Length, Inches	Diameter, Inches
15	2

Shell Cases

These, although they are not rolled (except canister shell cases) fall into this division, as they are made by the case rollers and consist of paper and paste.

There are two ways of making round shell cases. One, roughly speaking, consists of papering the inside of a ball whereas the other is papering the outside of a ball. The first

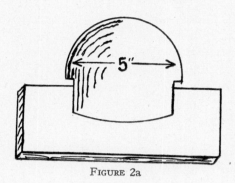

FIGURE 2a

Mold for Shell Cases

makes the nicer shell but requires more skill. We shall take a 6 inch shell as an example, since it is the most popular size and the same method is employed for all. We shall explain the second method first as it is the easier one. First, procure a mold as shown in sketch (Fig. 2a) 5 inches in diameter. Then cut strips of straw, wastepaper board or Kraft paper, etc., about ¾ inch wide and 4½ inches long and paste fifty, or more, onto a board, one on top of another, with so much paste between them that they become soft and pulpy rather than stick together. A red building paper sold in rolls makes a better case than any other kind. Two kinds of paper should be used and a separate stack of each kind prepared. The object of this is to make it easier for one to see where 1 layer of paper begins and the other ends.

Now, smear the ball or mold well with paste so that it will be wet enough to keep the paper from drying and sticking to it before the shell case can be finished. Then take strips of paper from one pile first and lay them on the mold, beginning on top and running half-way down the side. Lay the second strip so that it will lap over the first one about ¼ inch at the lower end and almost over it on top, but ½ inch lower down. The third strip should start ½ inch still further down, while the fourth strip again starts at the top. This will prevent the case from becoming egg shaped as a result of having too much paper in one place. Continue this until the entire upper half of the ball has been covered. Each strip must be pressed down firmly and all surplus paste squeezed out with the fingers. Now repeat the operation, using the other kind of paper.

Another way is to cut the strips a foot long and, after softening with paste as above, lay them on the mold from the top to the middle, tearing off the strip at the required point, letting the second and third pieces start $\frac{1}{2}$ inch below the other. In this manner it is possible to avoid having the top too thick. After the third layer has been put on, one should be laid crosswise, crossing as much of the first layers as possible. This process is continued, pressing each strip as firmly as possible until the case is about $\frac{3}{8}$ inch thick, while wet, at the edges where it is usually the thinnest and not over $\frac{5}{8}$ inch on the top. If the work has been properly done, the half shell can now be slipped off and allowed to dry. When dried the lower edge should be trimmed off with a sharp knife, at a point that will make a hemisphere. When the two halves are joined a sphere results.

The other method is to have a wooden block hollowed out so as to form a perfect hemisphere $5\frac{1}{2}$ inches in diameter. A mold may also be made by taking a ball of this size, oiling it well and setting it half-way in a box of wet plaster of Paris. Now then, proceed as before, but paste the strips inside of the hollow instead of on the outside of the ball. This will make a smoother-looking shell and a stronger one when properly done. The paper may conveniently be cut into strips a foot, or more, long and torn off as they reach the edge of the mold. In this way, all waste is avoided and the edge is even and regular. The strips should be pressed very firmly as the quality of the shell case depends on this. If the pressure against the fingers, in rubbing out the paste, makes them sore, a piece of wood about 3 inches long and $1\frac{1}{2}$ inches wide, rounded and slightly curved on one end, may be used as a presser. If the work has been done well the case, when dried, should be as firm as wood. To remove the wet shell case from the mold, first place on the bottom of it two strips of cloth at right angles, with the ends extending far enough over the sides to permit the pulling out of the completed case.

When the halves have been evenly trimmed place them together so as to form a sphere and secure the joint with a strip of canvas smeared with glue. Then put on 1 or 2 more layers of paper. After drying, bore a hole for a fuse through one end or, still better, bore a hole with a wood bit through one half from the inside before joining the halves.

Of course, paper shell cases may now be obtained ready-made from manufacturers of pulp products.

In addition to these methods, very good shells can be easily and quickly made with hollow balls of zinc, tin or wood. The wood half shells need only to be well glued together and they are ready for use. Those of zinc and tin require papering just as described for shell making with a round mold but the entire ball is papered until the paper is about ¼ inch thick for 6 inch and ½ inch for 10 inch shells.

Figure 2b shows, in outline, various parts of bombshells.

Beginning at the upper left is shown the manner of laying the folds of paper at the bottom of canister shells. The paper is softened with plenty of paste while the folds are pressed down firmly with the fingers, and surplus paste is squeezed out.

The top of the shell shows the manner of inserting the head. A hole for the fuse has been bored previously; the fuse is smeared well with glue and slipped through. Surplus glue appears around the top of the fuse. The bottom of the fuse should be empty for about ½ inch into which has been inserted a piece of quickmatch, secured with priming.

The center cut shows a complete shell with match, driving charge, etc. If the bottom and top are made of wood, the rabbet is well covered with glue when being inserted and is further secured by driving a few small wire nails through the case and into it.

The right-hand cut shows a round shell fused at the bottom which represents an alternate method. In this case blasting fuse is preferred. The hole through the shell case is bored so as to assure a close fit. Plenty of glue is used on both inside and outside as shown and a little twine wound around so as to form a shoulder. The fuse is attached to the lower half before the shell is loaded. Both ends of the fuse must be primed.

Canister or Cylindrical Shells

Cylindrical or canister shells are very popular with firework-makers because they are easily and quickly made and hold much more than round ones. On the other hand, a canister shell never breaks as beautifully as a round one. It blows out from the ends more like a large rocket. Cases for canister shells need no detailed description here as they are made in the same way as any other heavy round case. A former of the required size is procured and the case rolled thereon just as for a mine, any kind of heavy paper being satisfactory. When thoroughly dry the wooden heads are glued if fitted with a flange, and nailed with 1 inch wire nails; then they are carefully sealed all around with several thicknesses of strong paper. More detailed directions follow.

For a shell of 4 inch diameter take a wooden former $3\frac{1}{4}$ inches in diameter and 8 inches long, provided with a suitable handle. Procure good strong paper, Kraft, sheathing or rag board. Cut this across the grain in convenient lengths and of a width equal to the desired length of the shell. Paste it heavily with a good grade of thick flour paste and roll enough of it around the former to make a case $\frac{1}{4}$ inch thick. Do not allow paper to soften too much before rolling. Remove from former and dry in the shade.

Turn some pieces of maple (or similar wood) $\frac{3}{8}$ inch thick and $3\frac{3}{4}$ inches in outside diameter and $3\frac{1}{4}$ inches inside so as

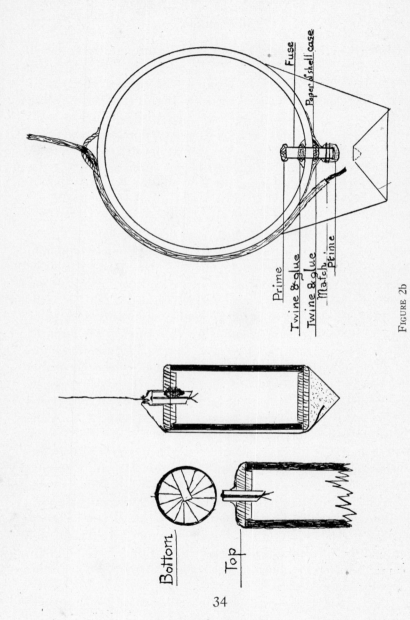

Fuse

Paper shell case

Prime

Twine & glue

Twine & glue

Match

Prime

Bottom

Top

FIGURE 2b

Parts of Bombshells

34

to fit snugly into the shell cases. Bore a hole into the top one for the fuse. Glue the bottom one into the case; fill the shell as desired; add bursting charge; glue the fuse in place with a surplus of glue on the inside and outside; and glue the top into the shell case.

When glue is set, cut some sheets of Kraft paper (24 inches by 36 inches—30 pound) across the grain, 4 inches longer than the shell, or a little more, so that the edges will lap when it is pasted into place. Soak this thoroughly with thin paste and roll about 5 layers around the shell, leaving an equal amount projecting at each end. With scissors, snip the projections into ½ inch strips longitudinally and work over top and bottom as neatly as possible so that they lie evenly and tightly around the fuse and there are no openings in the bottom. When dry it is matched and charged as described for round shells. This method is much faster than making round shells.

Another method which seems to be rather extensively used by Italians is described by Prof. Tenney L. Davis in his *Chemistry of Powder and Explosives*.[1]

For a 4 inch diameter shell, take a strip of bogus or news board, cut to the desired length, and roll tightly on a form without paste. When it is nearly all rolled, a strip of medium-weight Kraft paper, 4 inches wider than the other strip, is rolled in and around the tube several times and is pasted to hold it in position. Three circular disks of pasteboard of the same diameter as the bogus tube (3½ inches) are taken and a ⅝ inch hole is punched in the center of two of them. The fuse is inserted through the hole in one of them and glued heavily on the inside. When this is thoroughly dry, the disk is glued to one end of the bogus tube, the matched end of the fuse being outside. The outer wrapper of Kraft paper is folded over carefully onto the disk, glued and rubbed down smoothly. Then the second perforated disk is placed on top of it.

---

[1] T. L. Davis, *Chemistry of Powder and Explosives,* New York, John Wiley and Sons, 1943.

The shell case is now turned over, placed on bench with hole to receive fuse and filled. Bursting charge is added. A disk of pasteboard is placed over the stars and powder, pressed down against the end of the bogus body and glued. The outer Kraft paper wrapper is folded and glued over the end.

The shell case is dried and wound with strong jute twine. It is first wound lengthwise; the twine is wrapped as tightly as possible and as firmly against the fuse as may be. Each time that it passes the fuse the plane of winding is advanced by about 10° until 36 turns have been laid on. Then 36 turns are wound around the sides of the cylinder at right angles to the first winding. The shell is now ready to be *pasted in*. For this purpose, 50 pound Kraft paper is cut into strips of several dimensions, the length of the strips running across the grain of the paper. A strip of this paper is rubbed with paste until it is thoroughly impregnated. It is then laid on the bench and the shell rolled up in it. Stand shell upright, fuse end up, and tear the portion of wet Kraft paper which extends above the body of the shell into strips about ¾ inch wide. Rub down smoothly, permitting each to overlap the other on top of the shell and around the fuse about ½ inch. Reverse shell and repeat operation on bottom. Dry outdoors when propelling charge is attached.

A piece of piped match is laid alongside of the shell; both are rolled up without paste in four thicknesses of 30 pound Kraft paper, wide enough to extend 4 inches beyond the ends of the shell and held lightly in place by two strings near the ends of the case. Turn the bottom up; expose 3 inches of match and insert the second piece of the match in the pipe, tying with a string. Introduce blowing charge of 2F gunpowder; the 2 inner layers of Kraft paper are folded down upon it, pressed firmly and the outer layers pleated to center, tied and trimmed close to the string. Reverse; scrape the fuse clean in case it has been touched with paste. Two pieces of match are crossed over the end of the fuse, bent down alongside and tied in position. The piped match, which leads to the blowing charge, is now laid down upon the end of the cylinder, up to the end of the fuse tube, then bent up alongside of the fuse tube, then bent across its end and down the other side, and then bent back upon itself and tied in this position. Before it is tied a small hole is made in the match pipe where it passes the end of the Roman fuse and a piece of flat black match is inserted. The 2 inner layers of the Kraft paper are now pleated around the base of the fuse and tied close to the shell. The 2 outer layers are pleated and tied above the fuse, a 3 foot length of piped match extending from the upper end of the package. A few inches of black match are now bared and an extra piece of black match is inserted and tied in place by a string about 1 inch back from the end of the pipe and covered with a section of lance tube.

Drying Cases

For drying all cases more than 6 inches long, racks are most convenient. These are made of strips of ⅞ x 2 inches cypress or other light wood. When filled with cases they should be moved to a well-ventilated room or covered platform. If placed in the sun the cases will be badly warped in drying. The longer the cases, the farther apart the strips of the racks should be. Center strips for carrying racks and the end strips should be made of ⅞ x 3 inches wood (Fig. 3).

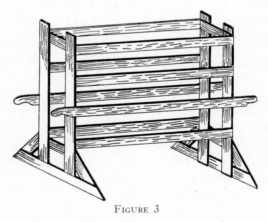

FIGURE 3

Drying Rack

Small cases may be thrown lightly into sieves 2 feet wide, 4 feet long and 4 inches deep, made of 1 inch material, the bottoms being covered with ½ inch mesh galvanized hardware cloth.

When cases are stored care should be taken to protect them from roaches and mice, as these are attracted by the paste.

Formers

All paper cases are rolled on formers of one kind or another. For rockets, gerbes, etc., these may consist of hardwood sticks but better ones are made of light brass tubing with an outside diameter equal to the inside diameter of the case desired. They should be 1 or 2 inches longer than the intended case and fitted with wood handles so that the former may be easily removed when the case is rolled (Fig. 4).

FIGURE 4

Rocket Case Former

Mines, etc., are rolled on wooden formers, the ends of which are turned down to convenient size to fit the hand. Roman candles are rolled on rods of machine steel; match pipes and pin wheels, on thin brass or steel rods; lances, on small brass tubing.

Paste

Paste is an essential which is in constant demand in every department of the factory. Ready-made and cold-water pastes are so easily obtained now that few persons care to bother making them themselves. However, for those who may not be in a position to obtain paste, the following standard method of manufacture is recommended.

Mix 4 ounces of wheat flour with 8 ounces of water and ⅛ ounce of powdered alum, stirring until there are no lumps. Pour this slowly, and with constant stirring, into 16 ounces of boiling water to which have been added 5 drops of carbolic acid, 5 drops of oil of cloves and 2 grains of corrosive sublimate. When cold it should be ready for use.

There is another method which is not only the simplest and best but requires no preservative and results in a product, which if made according to directions, will keep for a month or more in winter. According to this process, the batter is allowed to sour before heating and is heated by the addition of boiling water instead of being placed directly on the fire where it is likely to get lumpy or over-cooked. The following details are for making paste in lots of 3 or 4 buckets a day.

Procure two deep wooden tubs of about 20 gallons capacity. Buy a barrel of the cheapest grade of wheat flour. Sour and wormy flour will do. Put 2 or 3 bucketfuls of flour into one of the tubs and add water, stirring with a paddle until the water and flour are well mixed and thick enough to handle with convenience. It does not matter whether the mixture is lumpy, too thin or too thick, as all this is corrected later. The tub should be not more than one-third full. Allow it to rest in a warm place (about 90° F.) for 2 or 3 days. By this time fermentation will have started. When the fermentation is complete the flour will settle as a heavy batter on the bottom of the tub with a sour, brownish liquid over it. Pour this liquid off and fill several buckets about one-third full of the batter.

From a water boiler of 10 gallons capacity with a faucet in the bottom, fill the buckets with boiling water. While the water is running in stir the mixture briskly. The contents of the bucket will at first be as thin as milk but will gradually thicken as more water is added until they can be stirred only with difficulty. If all the details have been followed correctly, a bucket of clear, clean and very sticky paste, free from lumps, will result. The second tub may be used alternately with the first for souring batter while the contents of the first tub are being used for paste making. This paste, having been soured before cooking, cannot sour again and will not become watery. Also, no chemicals need be added to preserve it.

Glue and dextrin are sometimes added to make paste bind better (for finishing shells). The Japanese sandpaper their finished shells and rub them with paste while wet, then with whiting, finally varnishing them so that the finished shells are very smooth. This, of course, is unnecessary when shells are intended for prompt use.

Crimping

Sometimes, to eliminate the use of clay, gerbes and the like are choked or crimped to reduce the opening. This is so particularly in the case of serpents, saucissons and such small cases. One-, two- and three-ounce rocket cases are crimped and the hollow center is made by driving a spindle into them after ramming, as explained later. Crimping is done in two ways, either by hand or by machine as shown in Fig. 5.

Hand-crimping is done by taking a turn of strong string or piano wire around the case about ¾ inch from the end, while it is still wet, and drawing tightly while turning the case slightly so as to make a neat job. One end of the string or wire should be fastened to a wall or some unyielding object while the other is passed around the body (a, Fig. 5). A nipple with a short point, slightly smaller in diameter than the desired opening to be left in the case, should be inserted before the string is drawn so that the end of the case will be kept open and crimping neatly done (b, Fig. 5). A mechanical device, made by a Cincinnati machine works, does the work very neatly and much more quickly than the string process (c, Fig. 5).

Ramming

As this operation will be described in detail under each

individual article, only a few general directions will be given here.

All ramming should be done in small sheds, as far removed from the rest of the factory as possible, and one side should

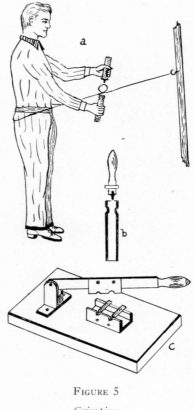

FIGURE 5

Crimping

always be open. The operator should have his back toward this side while at work. A stout wooden block, resting either on the ground or over a foundation, should be used for ramming. For heavy work the best mallets are those made of rawhide. These are round and range from ½ to 10 pounds in weight. A good weight for average work is about 2 pounds.

Ramming tools and nipples should be made of gun metal or brass, whereas the spindles for rockets must be made of steel.

Scoops for taking up the required amount of composition at one time can be made of tin or any light metal and should be provided in different sizes, from about ½ inch wide and 1 inch

FIGURE 6

Ramming

long to 1 inch across and 3 inches long with about six intermediate sizes. Some compositions work better when rammed in small quantities.

Ramming with Rod and Funnel

For small work such as serpents, saucissons, etc., make a funnel about 4 inches high, having a 3 inch diameter on top, ⁵⁄₁₆ inch at the bottom, without a spout (a, Fig. 7).

Procure a rod ¼ inch in diameter and 12 inches long, or longer, according to the work to be done. A wooden knob may be attached to the top of the rod for convenience in ramming (c, Fig. 7). In use a case is slipped on a nipple (d, Fig. 7). The small end of the funnel, which is half full of composition, is inserted in the top of a case and, with the rod moved up and down, striking the bottom firmly each time, the composition becomes rammed with sufficient solidity. When a case has been rammed to within ½ inch of the top, the funnel is removed and

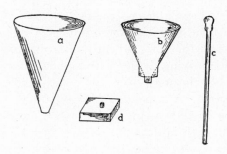

FIGURE 7

Ramming Implements

a charge of clay is added. Strike the clay a few blows with a light mallet and suitable drift or rammer. Sometimes the ends are closed after the last charge of composition. About half of the end of the case, sheet by sheet is bent in with a small suitable punch, somewhat as is done with Chinese crackers. A small amount of moist clay should be applied before the bending in.

The arrangement for making lances is somewhat lighter. The funnel (b, Fig. 7) is very efficient. It is 2½ inches on top and 2½ inches high with a ¾ inch shoulder on bottom and a spout of ¼ inch outside diameter, projecting from the bottom for ¼ inch. This, when removed from the lance, leaves just the proper space in the case empty for priming.

Matching

This is the term used to designate that operation in pyrotechny which consists of bringing fire to the various parts of devices to be burned. In most of the individual articles a short piece of match is twisted in the nosing of the wrapper or fastened otherwise. In set pieces this operation takes on a great importance.

Matching of lancework is fully described under that heading. In the case of set pieces consisting of gerbes, wheels, etc., the gerbe is first primed by smearing a little priming on the inside of the choked end of the case. A nosing is put on, consisting of 2 or 3 turns of strong paper, rolled around the case so as to project 1½ to 2 inches beyond the end, depending on the size of the case. About ½ inch of the piping is removed from a length of quickmatch. This is bent back, inserted into the nosing and secured by tying tightly with two half hitches of twine. The match is now brought over to the next gerbe and bent at right angles over it. At a point, 2 inches from there, it is again bent back onto itself to point of last bend, and bent again at right angles so as to lead to the next gerbe (Fig. 8). At the bottom of this bend the piping is cut off with a sharp knife, baring the match, and this portion is shoved into the nosing of a second gerbe and secured by tying as before (Fig. 8). Candles, wheel cases, etc., are treated in the same manner.

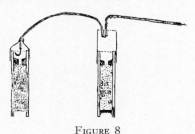

FIGURE 8

Matching Gerbes

If a gerbe has been properly primed it is not necessary for the match to enter the choke as the fire will reach it from priming.

It is a good plan to have the leader, from which a piece of fireworks is ignited, run to each section of same, irrespective of the fact that those sections are already connected to one another in the process of matching it. Sometimes a section of match has a defect and will go out in the center of the pipe. Therefore, it is advisable to have the match joined wherever it crosses, as for instance on top of a lance, guarding against failure to ignite.

If it is desired to have one part of a piece burn after the other part has been burning, as when candles and gerbes are used in connection with lance work, these gerbes, etc., are matched to a separate leader which may be fired by hand after the lances are half consumed, or the second leader may be connected to several lances about half-way down so that when they have burned this far the balance will be lighted automatically. This is done because lances burn so much longer than candles or gerbes. If fired all at once, the gerbes, etc., would be finished before the lances are half burned. As it is most effective to have a display of any kind increase as it develops, with the climax at the end, it is best, in a device composed of lances, gerbes and candles, to start with the lances. When these are one-third burned, start the candles and, when they are two-thirds finished, fire the gerbes as a finalé.

Priming

In order to insure lighting, especially in exhibition work, all gerbes, wheel cases, lances, etc., are primed. This consists of smearing a little moist gun powder about the mouth of the case. Priming is made by adding water to grain or meal powder until the powder becomes pasty. A little alcohol and dextrin may be added to advantage.

Finishing

In factories where stock or shelf goods are made, the finishing department is quite important. All kinds of store fireworks are covered with some kind of colored paper and often stripes and borders are added. This is especially so with large mines and 4 to 8 pound rockets. Candles, serpents, small mines, rockets and triangles are covered with vari-colored poster paper. Flowerpots, floral shells and fountains are generally covered with calico or glazed papers, stripes being added where desired. The cuts of finishing paper for candles and rockets are for use with articles of the sizes described in this book. The cuts are usually 2 inches longer than the articles to be papered if they are to be matched at one end and tucked in at the other, 1 inch longer where matching only is done, and the same length where only the case is covered as in mines, etc.

Sky Rockets

		Size, Inches
1 oz.		3 x 5
2 "		3 x 6
3 "		4 x 6½
	With Cone	
4 "		4 x 6½
6 "		5 x 7
8 "		6 x 8½
1 lb.		6 x 9½
2 "		7 x 10½

Roman Candles

	Size, Inches
1 ball	2 x 6
2 "	2 x 7
3 "	3 x 8
4 "	3 x 10
6 "	4 x 14½
8 "	4 x 17
10 "	5 x 19
12 "	5 x 21
15 "	6 x 24
20 "	6 x 28
25 "	6 x 34
30 "	6 x 38

Wrapping

It is as important and as difficult to learn to make a good, neat, tight and strong bundle as to learn any other part of the stock fireworks business.

Roman candles with one to four balls are packed 3 dozen in a bundle; from six balls up, 1 dozen in a package. The packages of 1 dozen are made in two forms, i.e., four- and six-sided (Fig. 9, b and a, respectively).

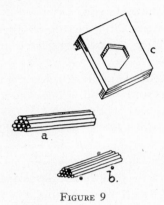

FIGURE 9

Bundles of Roman Candles

To make the four-sided package of 1 dozen eight ball candles lay five of them on the bench. Mark the space they cover and fit into the top of the bench four wood pins, 1 inch of which projects above (two on each long side of the space occupied by the five candles, so that the candles may be easily laid between them). Now cut a sheet of Manila paper 19 inches long and 14 inches wide and place this between the pins just as the candles lay before. Replace the five candles on the paper between the pins. On top of these five place four more candles and on top of them place three (this makes twelve). Draw the paper tightly over them and fold it as in wrapping a bundle.

Close the ends as follows: with two fingers press the top of the folded paper over the end of the three top candles; then, holding them down with both thumbs, fold in the two sides of package with the first and second fingers of each hand at the same time. Then holding these folds with the left hand, lift the opposite end of the bundle with the right hand. This will cause the bottom to fold itself over the other folds. With a brush dipped in thick paste give the end a daub on the last fold and while the bundle is still standing on this end fold the top end the same way. Before turning in the last fold give it a daub of paste as you did at the other end. Lean against the wall and place a paper weight or small tile on top of the end to hold it in place until dry. After the packages have dried, the labels are affixed.

To make the six-sided bundles (a, Fig. 9) one must first learn to form the candles in the hand. Count out a dozen candles and encircle the bunch with both hands. Work them about until they form the six-sided bunch with three candles on the three wide sides and two on the other three sides. Lay them on the wrapping paper (cut as described above, but preferably wider) holding them tightly so they retain the proper form. It is difficult to describe the method of getting the paper around them without having them fall in a heap. This technic is difficult to master, although easy enough when learned. Once the bundle is on the wrapping sheet one hand is sufficient to maintain its form. With the other hand lift the side of the sheet nearest to you and bend it partly around the package. Hold it while the other hand is released long enough to take the paper up on that side. Straighten and flatten it well on over the candles and roll up the bundle. Paste this edge and lap it on the bundle and then you are ready for the corners.

If the bundle has been properly made, the top row must consist of two candles; the second row, three; the third row, four and the bottom row, three. Now with two fingers of the left hand bend the paper down from the top over the two top candles first; then bend in the two upper sides; then the two lower sides. Finally, by lifting the bundle from one end the bottom folds itself over all the others while the top is folded in the same way. A little paste secures it, as described above.

The bundles of smaller candles are formed in a wooden form, hollowed out to the size of a bundle of 3 dozen (c, Fig. 9) and when it is filled with the required number the candles are tied with a string before wrapping.

Short stick rockets are neatly packed in paper boxes. Long sticks are packed as follows: Cut 18 gauge annealed iron wire into 6 inch pieces. Take six or a dozen rockets with the heads all even and work them in the hands until they form as square a bundle as possible, then bend one of the pieces of wire around the sticks just below the matches. This should be done with one hand while the other holds the bundle in shape. Now pass another wire around the sticks about a foot from the bottom. Cut some pieces of strawboard as wide as the bundle of rockets on the wide sides and long enough to go completely over the heads and down the other side nearly to the matches. Cut some wrapping paper 6 inches wider than the rocket head bunch and long enough to go twice around it. Paste the far edge for about 1 inch and lay the bundle of rockets, with the strawboard around it, on the sheet and wrap it up as tightly as possible. Fold in the upper end in the regular way; secure with a little paste and set aside, heads downward, to dry. Later, the other end may be gathered in and secured with another piece of wire.

Wheels, tourbillions, etc., are made into any desired kind of a bundle, while mines, fountains, etc., are given 1 or 2 turns of paper over the finishing to keep them clean. Serpents, flower-pots and torches are packed like Roman candles. Blue lights are packed similarly, usually a gross to the bundle. Large fancy rockets are packed heads and sticks separately, the heads in boxes and the sticks loose.

Wiring

For most purposes annealed iron wire, 18 to 20 gauge, is serviceable. The easiest and quickest way to use it for wiring rockets, triangles, etc., is to cut it into lengths of from 4 to 6 inches, according to the size of the work to be done. A large quantity can be cut at once by using a bench shear and cutting several hundred at a time. Rocket sticks can be quite securely fastened with one wire if a *gum board* (Fig. 10) is used; otherwise two wires are necessary. A gum board is made by taking a ½ inch board 6 inches long and nailing pieces of rocket stick around it on three sides on top, and one side on the bottom as shown. Put into this about 1 ounce of dextrin mixed with water to the consistency of jelly. Now it is ready for use.

FIGURE 10

Gum Board

Put a pile of rocket heads and wire to your left, and a bundle of sticks and the gum board to your right. Rub one side of a stick against the bottom of the gum board so that a little of the gum will adhere to it. Lay it with the gummed side against the rocket about three-quarters of the way to the cone. Hold it in this way in the left hand and, with the right hand, bend a wire around it, at about the middle, giving 1 turn on the side of the stick. Now, with a pair of nippers, give about 3 more turns, cutting the wire with the last turn. If no gum is used two wires are necessary.

Tying

In doing exhibition work, string plays a very important part and the best and most convenient knot for all purposes is the sailors' clove-hitch (Fig. 11).

FIGURE 11

Sailors' Clove-hitch

This is somewhat difficult to learn. The best way is to practice on a stick. Pass the string under the stick bringing the end
to the left of the loop; bring it over again to the left of the second loop and pass the end between the second and the first. An
ordinary tie of the free ends now secures the knot permanently.
This knot will be found invaluable in matching and many other
operations.

Labeling

This very easy operation may be simplified still further if it is
done in the right way. Take a board about a foot square. Smear
it well on top with thin paste and lay a label, face down, on it.
Cover this well with paste and place another label on top of it,
repeating the pasting and putting down of labels until several
dozens are on the board. This will soften them so that when
they are taken up and pressed with the fingers or paste brush
against the bundle to be labeled they will adhere firmly and lay
flat.

Designing

When it is desired to produce a portrait, a picture of a building, monument, etc., or a line of lettering in fireworks, the design is drawn first on a suitable floor with a piece of chalk fixed
into the end of a stick so that the designer may walk about
sketching his picture from the miniature plan as he walks along.
First, the floor is laid off into squares with a chalk line, 1 foot
each way and in multiples of 50 square feet, 5 feet wide and 10
feet long. For instance, if a picture 10 feet high and 20 feet
wide is desired it is composed of four sections 5 x 10 feet or two
high and two wide.

The sketch is now marked off with a ruler and divided into
200 equal squares, 10 high and 20 wide, corresponding to the
full-sized squares on the floor. These are numbered along
the edge of the sketch along the top and on one side. The

squares on the floor are numbered in the same way. With the chalk draw into each square on the floor the same lines as appear in the corresponding square of the sketch. When this is done, an exact reproduction of the small picture will be ready to be placed on the frames.

For lettering or lines of words this is not necessary as design, of the desired size and without enlargement, can usually be drawn directly onto the floor with free hand.

Part III

PRODUCTS OF MANUFACTURE

Match
(Quick Match)

THIS IS used for conveying fire to the combustible portion of pyrotechnical devices and is distinguished from a fuse by the fact that its effect is almost instantaneous, whereas a fuse burns at a comparatively slow and exact rate. It consists of cotton wicking impregnated with gunpowder and covered with a loose paper piping. As almost every piece of fireworks requires match for lighting and since lance-work and exhibition pieces are absolutely dependent on a good match for their successful operation, it is essential to make this very necessary article as nearly perfect as possible. There are several ways of making match which will be classified as the *French system,* the *English system* and *candle and rocket match manufacture.*

French System

Secure two pieces of 1 x 3 inch lumber and into one edge of each drive a number of six or eight penny wire nails (for half their length) about 1 inch apart. Set these pieces up horizontally, with the nail edges uppermost, about 3 feet above the ground, one at each end of a dry shed about 30 feet long. Wicking of the proper thickness can be secured in balls as desired. If not, take a number of balls of ordinary cotton wrapping twine (four to six, according to their thickness) and place them in a box with the ends tied together so as to make a single cord about $\frac{1}{8}$ inch thick. If you can secure the required thick-

53

ness take a ball of not less than twenty-four strands and fasten the end to the nail nearest the wall (on one of the above pieces of 1 x 3 inches) with the ball (or balls) of cord in a light box, so that it can unwind easily. Walk to the other end of the shed and, drawing it rather tightly, fasten it to the corresponding nail in the other strip of 1 x 3 inches by winding a few turns around the nail. Leave the box with the ball temporarily at this end of the shed.

In an agate pan mix 3 pounds of rifle powder thoroughly with 4 ounces of dextrin or gum arabic. Add water and stir with the fingers until all the grains are wet. Allow to stand for a few minutes until a small lump pressed between the fingers feels perfectly smooth and contains no more grains. Stir in some more water and a little alcohol until the mixture is about the consistency of mush.

Holding the pan in the left hand, take up a handful of the powder mixture under the first length of cord and work it well into the cotton. Hold the pan so as to catch any drippings and walk backward to the other end of shed, rubbing the powder into the cord as you go. When the end of the strand is reached go back to the beginning. Take some of the powder mixture in the right hand, pass the cord over the first joint of the first finger, place the thumb on top of it and again walk backward toward the other end, but without working any more powder into the cord. Simply allow it to run through the finger and thumb for the purpose of rubbing off rough, uneven places and leaving a smooth well-finished surface.

Now take up the box of cord again and pass it around the third or fourth nail to the right. Stretch a second length to the starting point, fastening it several nails away from the first strand. Proceed as with the first length, not touching the finished strand while working the powder into the second one. When it is finished move it to the second nail, stretching it tightly. Repeat with the following strands until all the powder is used up.

If the weather is dry, the match will be ready for piping in a day or two. In dry climates gum arabic makes a better match than dextrin but where there is much moisture in the air dextrin should be used. When the match is dry and stiff it may be cut and pipes threaded on.

Match pipes are made of 20 pound Manila or Kraft paper 24 x 36 inches. Cut strips 4 inches wide and 36 inches long and roll on a ¼ inch steel rod. Paste only the edge of the sheet on about 1 inch width. When piping the match, crease or gather the end of the first pipe, when in place, so that the next pipe may be slipped over it for about 1 inch.

A simple and clean method of making exhibition match and one which, to the best of the author's knowledge, is original, is as follows:

Make a cup of brass, about 3 inches in diameter on top, 2½ inches on the bottom and 2¼ inches deep as shown in Fig. 11a.

At the bottom attach a spout pointing upwards and terminating in an opening ⅛ to ³⁄₁₆ inch in diameter. A small dishpan, a 2 quart pudding pan and a regular match frame (Fig. 12) complete the apparatus. Then prepare the following formulas:

Formula 1		*Formula 2*	
Dextrin	3 oz.	Dextrin	1½ oz.
Gun powder	2 lb.	Gun powder	1 lb.
Water	2 pt.	Water	12 oz.
Alcohol	4 oz.		

In the dishpan place Formula 1, mixing the gun powder thoroughly with the dextrin, then add the water. When the powder is completely dispersed add alcohol and stir well. Into this unwind about 2 pounds of cotton wicking of not less than 24 to 30 strands and with a stick press it well into the powder mixture. In the pudding pan place Formula 2, proceeding as for Formula 1. This, however, should be thicker, like a soft putty.

Now take the end of the cord from the large dishpan and pass it through the spout of the brass can, from the inside. Fill the brass cup with Formula 2 and pull enough cord through

FIGURE 11a

Brass Cup for Making Exhibition Match

the spout so that the end may be attached to the match frame. Hold the cup in the left hand and revolve the frame with the right, placing the dishpan so that the cord will slide over a notch in the rim of the cup and through the powder mixture onto the frame. Separate each cord by about ¾ inch. If the spout of the cup fits the cord snugly a perfectly round, smooth match will result and if rubber gloves are worn the hands will not be soiled. Be sure to keep the small cup full.

English System

Make a light frame of wood like the frame of a mirror (Fig. 12) 6 feet long and 4 feet wide and hang it in an upright stand so that it can revolve like the mirror of a dresser. Then get a length of 24 strand cotton wicking, and unwind it into a pan 10 inches in diameter and 6 inches deep. In another similar pan put 2½ pounds of rifle powder and 4 ounces of dextrin. Mix well and cover with 2½ pints of water, stirring occasionally until the powder is dispersed. Then add 2½ ounces

FIGURE 12
Match Frame

of alcohol and mix again. Pour this over the wicking in the first pan taking care to leave the end of the wick hanging over the edge so that it can be found easily.

Beginning with this end run all the wicking into the empty pan, taking care to see that every part of it is well soaked with the wet powder, a little of which should remain after the cotton is removed for the first time. This may be poured over the pile of wet wick, pressing or kneading it so as to soak every part thoroughly. Now return it to the first pan as before, and it is ready for the frame. Tie the end of the wick to one end of the frame and let someone turn it slowly while feeding the match onto it with the cords about ¾ inch apart. When all of it is on the frame, remove it to a part of the floor covered by large sheets of paper. Support the frame on four blocks about 3 inches high, one at each corner of the frame.

Take a small sieve filled with meal powder and dust the meal powder carefully over the match so as to cover it evenly. Place the frame with the match in the sun or some other warm place to dry. Matches made by this process are all of one length, i.e., 2 yards, and are round in appearance. They burn fiercely but will not stand as severe usage as the matches previously described. They also take longer to make.

The following formula for exhibition match was used by the English pyrotechnist Robert Bruce.

Water	1	qt.
Gum arabic	5	oz.
Wood alcohol	½	pt.

Mix well and add meal powder until a mass about as thick as porridge or molasses is obtained. Soak the cotton wicking in this mixture for 3 hours; run onto frame and dust lightly with meal powder.

Match piping serves the double purpose of protecting the match from injury as well as making it burn faster. A piece 20 feet long will flash from one end to the other in less than a second.

Rocket and Candle Match Manufacture

Match made by either of the previous methods is too expensive for use with the cheap grade of stock fireworks on the market. Therefore, a simpler method is used for this purpose. In principle, it is essentially like the English method.

	Formula 1	Formula 2	Formula 3
Saltpeter	16	36	72
Fine charcoal	5	12	15
Sulfur	2½	6	10
Dextrin		1	2

In a small tub put a gallon of well-boiled starch and stir into it 15 pounds of a thoroughly mixed composition of Formula 1. Soak cotton wick of about 5 strand in this until nearly all of the composition is absorbed except a 1 inch layer which should still cover the cotton in the tub. Work it in well and run it onto a frame as directed in the preceding description.

The frame may be smaller for convenience in handling by one person as long lengths of match are not required. Neither does it need to be dusted with meal powder. If made well, however, it will burn freely and serve its purpose completely. When dry it is cut from the frame and tied in bunches 1 to 2 inches in diameter and cut into the desired lengths with a tobacco cutter or a large, sharp knife. No starch is needed when using Formulas 2 and 3.

When matching Roman candles, rockets and other shelf goods, where a large number of short pieces are required, these may be cut, several hundred at a time, by the use of a *tobacco cutter,* i.e., the implement used in country stores for cutting plug tobacco. The pieces are bunched into bundles about 1½ inches in diameter and pushed under the blade of the cutter. If the knife is sharp it will pass through the bundles of match easily.

Fuse
(*Blasting*)

This is used in pyrotechny in the production of cannon crackers and, to a lesser degree, in small bombshells. It consists essentially of a cotton tube containing finely divided gunpowder and burns at a rate of approximately 1 inch in 3 seconds. It can be had in sizes varying from ⅛ to ¼ inch in diameter and from the cheapest painted cotton kind to one heavily coated with gutta-percha for under-water work.

It is made by a very ingenious machine which weaves a cotton fabric around a small tube. As this tube is withdrawn its place is taken by the powder which is forced in through the opening of the tube. The largest factories are in Simmsboro, Conn.

Tableau Fire

This is about the simplest form of fireworks in use. It is made by thoroughly mixing the necessary ingredients to produce the desired color and heaping the mixture on an iron plate or board, so that it may be easily lighted. Or it may be put up in tin cans for the trade. Good tableau fire should burn brightly without sputtering and without excessive smoke. It should take fire easily but be free of the danger of spontaneous combustion. Lithographed cans are sometimes used as containers, designating, by their color, the color of the fire they contain. Firing instructions are printed on them. A small piece of match placed in each facilitates lighting.

White Fire

	Formula 1	Formula 2	Formula 3	Formula 4
Saltpeter	3	12	8	7
Sulfur	1	1	1	1
Metallic antimony	1			
Antimony sulfide	1	1		
Realgar			1	1½

Blue Fire

	Formula 1	Formula 2	Formula 3	Formula 4
Potassium perchlorate				24
Potassium chlorate	6	16	8	
Paris green	4		6	
Shellac			½	
Stearin	1		1	2
Barium nitrate	4		7	
Calomel		12	1	
Sal ammoniac	1			
Copper ammonium chloride		4		6
Asphaltum				1
Lactose		6		

In working with mixtures containing Paris green it should be remembered that this substance is very poisonous and a handkerchief should be tied over the nose before handling it.

Red Fire

	Formula 1	Formula 2	Formula 3	Formula 4
Strontium nitrate	80	10	16	14
Potassium chlorate	20	4	8	4
Shellac			3	
Red or Kauri gum	12	3		
Asphaltum				3
Charcoal		1		
Dextrin	1			
Fine sawdust	12			
Rosin		1		
Lampblack	1			

If one desires to increase the bulk of tableau fires, 10 to 30% of fine sawdust may be added to any of the several formulas without materially affecting the color. If the sawdust will not pass freely through the sieve it may be added after the other ingredients are sifted, and mixed and rubbed in with the hands.

PINK FIRE

	Formula 1	Formula 2	Formula 3	Formula 4
Strontium nitrate	48	16	18	
Saltpeter	12	4	7	
Potassium perchlorate				16
Sulfur	5	2	2	
Plaster of Paris				4
Shellac				3
Charcoal	4	1	½	
Red gum		3	2	
Dextrin			½	

The first three of these should not cost over 7¢ per pound; whereas the red mixings run about 9½¢ per pound. The fourth mixing is very beautiful although somewhat more expensive and smoky. Lime is sometimes used in pink fires, but it requires potassium chlorate and is more smoky.

YELLOW FIRE

Barium nitrate	36
Sodium oxalate	6
Sulfur	3
Red gum	5

This mixing is both good and cheap.

GREEN FIRE

	Formula 1	Formula 2	Formula 3
Barium nitrate	8	9	4
Potassium chlorate	4	3	2
Shellac		1	1½
Red gum or K. D. Dust	2		
Dextrin		1/16	
Fine sawdust		½	
Sal ammoniac	1		

SMOKELESS TABLEAU FIRES

Red

Strontium nitrate	8
Picric acid	5
Charcoal	2
Shellac	1

Green

Barium nitrate	4
Picric acid	2
Charcoal	1

Dissolve picric acid in the smallest possible amount of boiling water; add strontia or baryta; stir until cold and dry on a filter or a piece of cloth. Use as little water as possible even if all the picric acid is not dissolved.

It should be observed in all mixings, that the formulas cannot be considered absolute as the purity and general characteristics of the chemicals differ. All mixtures must be tested and regulated to the existing conditions of materials, climate, etc. If tableau fire burns too slowly, add more potassium or charcoal; if too fast, add more strontium or barium, etc. In rocket, candle or gerbe compositions, saltpeter or meal powder will increase the combustion whereas coal and sulfur will retard it.

Torches

Torches may be classified according to the purposes for which they are intended. Military torches or *flares* have but one requirement, i.e., that they produce the maximum illumina-

FIGURE 13

Fusee

tion of the deepest hue of color desired. Since these are fully described in special works issued by the government and really form no part of commercial fireworks, it will not be necessary to devote further space to them here. Railway torches or *fusees* are the cheapest form of red light. Any signal capable of attracting the attention of the engineer is all that is required. The laws requiring motor vehicles to carry fusees for use in case of forced stops along the highway has added greatly to the demand for this item. Fusees are usually ¾ inch in diameter and 8 to

12 inches long, exclusive of the handle, and burn from 5 to 30 minutes.

	Formula 1	Formula 2 *	Formula 3 [2]
Potassium perchlorate		4	15
Strontium nitrate	16	40	132
Saltpeter	4		
Sulfur	5	5	25
Fine charcoal	1		
Red gum		2	
Sawdust	1		20
Petrolatum		1	

* Moisten with kerosene before ramming.

Fusees are provided with a slip cap which is used for igniting them. The top of the torch is covered with cloth which is painted with a mixture of:

Potassium chlorate	6
Antimony sulfide	2
Glue	1

The end of the cap is similarly painted with:

Black oxide of manganese	8
Amorphous phosphorus	10
Glue	3

When the cap is pulled off and struck against the end of the fusee it takes fire like a safety match. With some compositions it is necessary to have a little starting fire at the top of the torch just under the capping or priming which will insure easy ignition. After the cap is in place it is covered with a strip of friction tape to protect the striking composition. The ends of this strip are brought down about ½ inch below the end of the cap and secured by a strip of paper pasted around to hold it.

Parade Torches

In making parade torches for campaign purposes, a cheap fire suffices and competition induces the manufacturer to produce

[2] According to Clark. T. L. Davis, *Chemistry of Powder and Explosives,* New York, John Wiley and Sons, 1943.

the largest article for the lowest price. One method in use is to add 50% of fine sawdust to the mixing. This does not greatly affect the burning of the torch and makes it look twice as large at practically no increase in cost. The following is a good formula:

Strontium nitrate	30
Potassium chlorate	8
Red gum	7

Sawdust may be added. The torches are usually $\frac{3}{4}$ inch in diameter, 12 inches long and should burn 8 to 10 minutes. All torches should be rammed as hard as possible. One of the large manufacturers has a ramming machine similar in principle to the candle-ramming machine described later, but of heavier construction and ramming twenty-four torches at a time. It is simple and works quite satisfactorily.

A cheap method of ramming campaign torches where price competition is very keen is to moisten the composition with dilute dextrin solution until it is damp enough to hold together when a handful is tightly squeezed. A dozen torch cases are tied in a bundle and pressed into a pile of this composition on a slab. It is then moved to a clear part of the slab and jolted a few times. More composition is added, the jarring repeated,

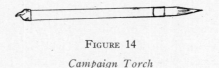

FIGURE 14

Campaign Torch

and this is continued until the torches are full. They are then set aside to dry. In this manner a dozen torches may be rammed in a minute without a machine. The handles are attached with a strip of gummed paper, 2 inches wide, half of which encircles the torch and the other half the end of the torch handle. The top is nosed and matched in the regular way (Fig. 14).

A better way of ramming torches by hand is as follows: Tie

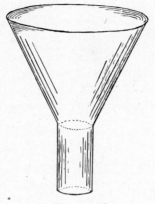

FIGURE 15

Funnel for Ramming Torches

the cases in bundles of twelve; place them on the ramming block and insert the spout of the funnel (Fig. 15) into one of them. Then pass a suitable rod about ¼ inch smaller in diameter than the funnel spout through it until it also rests on the block. Now, with a scoop fill the funnel with composition. Steadying the funnel with the left hand, grasp the rod firmly with the right hand, raising it about 6 inches. Drive it with a firm stroke up and down as the composition runs into the case. Continue this operation until the case is full. When the funnel is removed, the space occupied by the spout will remain. Insert the handle in this space using a little gum or glue.

Torches for Mardi Gras Carnival Parades

Torches for this purpose present the most exacting requirements. The following formulas are the result of more than 30 years of experimenting. Some exceptional mixtures, as well as beautiful colors, have been developed, which, in modified form, may be used for box stars, shells, etc.

Carnival parade torches must be of deep color, give maximum illumination, burn slowly and cleanly; they must not be expensive and should give off very little smoke. The entire elimination of smoke is not desirable, however, as this forms a background, making the tableaux stand out more. Still, it must not be enough to choke the maskers. The torches should burn approximately 15 minutes so that the carriers are not over-burdened with excessive weight during a two-hour march. The size is ⅞ inch in diameter and 18 inches long exclusive of handle.

RED

The following formulas are all good:

	Formula 1	Formula 2	Formula 3	Formula 4
Strontium nitrate	16	14	40	9
Potassium chlorate	8	4	8	
Potassium perchlorate				2
Shellac	3			
Red gum			7½	1
Asphaltum		3		
Sulfur, coarsely ground				2

PINK

The formula given under *box stars* results in a beautiful pastel shade but also much smoke. All mixings containing calcium compounds have this defect because of the light, white and abundant nature of their products of combustion.

ORANGE *

Strontium nitrate	36
Sodium oxalate	8
Shellac	5
Sulfur	3
Potassium perchlorate	5

YELLOW **

Barium nitrate	7
Potassium perchlorate	2
Sodium oxalate	1
Sulfur	½
Red gum	1

* This torch will deteriorate in a few months and become useless.
** Cryolite may also be used for producing yellow fires and torches and is not hygroscopic.

GREEN

	Formula 1	Formula 2	Formula 3	Formula 4 ***
Barium chlorate	25			
Barium nitrate	60	40	30	40
Potassium chlorate	5	11		12
Potassium perchlorate			6	
K. D. gum		6	2	
Sulfur			3	
Sal ammoniac		1		
Shellac	10			4
Stearin	1			
Petrolatum				1

BLUE

	Formula 1	Formula 2	Formula 3
Potassium perchlorate	5	24	24
Paris green	2		
Copper ammonium sulfate		6	
Copper ammonium chloride			6
Dextrin	1		
Calomel	1		
Sugar of milk		2	
Sulfur		9	
Stearin			2
Asphaltum			1

PURPLE

Strontium nitrate	7
Potassium perchlorate	9
Black copper sulfide, fused	6
Calomel	3
Sulfur, ground	5

The orange and purple torches are exceptionally beautiful and have been used very effectively in Mardi Gras parades in New Orleans, the Rex colors being purple, green and gold. Great care must be observed in mixing compounds containing sodium

*** Sift barium nitrate and mix with the petrolatum thoroughly. Sift potassium chlorate and mix with shellac. Add barium and petrolatum; mix and sift twice. Ram into cases ¾ inch in diameter and 12 inches long exclusive of handle socket. This will burn 10 minutes with a fairly good color but not equal to that of formula No. 1. However, its cost is considerably lower.

oxalate. All the ingredients must be perfectly dry and in damp climates it is best to mix only on clear days. Use waxed paper for the torch cases. The smallest amount of moisture will cause the oxalate to decompose forming with other ingredients of the mixture sodium nitrate or sodium chloride which are hygroscopic. The torches soon become so wet that they will not burn. Torches should be packed in moistureproof containers immediately after manufacture.

In cutting the paper for a 15 minute torch of $\frac{7}{8}$ inch diameter and 18 inches long, cut 35 to 40 pound Kraft paper so (that it will roll with the grain) 18 inches in length and across the grain $11\frac{3}{4}$ inches. This will give 4 complete turns and cause more regular burning. Pasting the outer edge on 3 to 4 inches will be sufficient. However, if both edges are pasted a stiffer case will result.

Capping and Matching Parade Torches

A good method of doing this is as follows: Cut cotton cloth into pieces about 2 inches square; cover them with paste and bend them securely over the top of the torch as shown in Fig.

Priming

FIGURE 16

Torch with Match

16. When they have dried, punch a hole about 1 inch deep through the cloth and into the top of the torch, with an awl about ⅛ inch in diameter, into which the match is inserted. Then make up some thin priming of gun powder, gum water and a little alcohol. If this is of the proper consistency it will be easy to place a little on top of the torch, with a priming stick, so as to secure the match in place as shown. This assists ignition and prevents the match from being pulled out.

Aluminum Torches

This beautiful piece of pyrotechny was first introduced into parades by the author, with sensational results, about 35 years ago. Twelve men were placed at the head of the line of march, each one with a burning aluminum torch. The effect of an avalanche of fire was produced. For this torch, a case ½ inch in diameter and 16 inches long is used, with a round wooden handle 6 inches long. They are rammed and matched much as other torches are. The following are good formulas for the mixing:

	Formula 1	Formula 2 *
Potassium perchlorate	13	8
Fine aluminum powder	6	3
Flake aluminum	5	
Dextrin or lycopodium	1	
Flour		1

Moisten slightly with alcohol before ramming. A beautiful modication of this is the red and aluminum torch.

RED AND ALUMINUM TORCH

	Formula 1	Formula 2
Strontium nitrate	35	36
Potassium perchlorate	7	
Shellac	4	1
Flake aluminum	4	
Lycopodium	1	
Sulfur		2
Mixed aluminum		9

* A. St. H. Brock. Personal Communication.

These torches should be ⅞ inch in diameter and 18 inches long. Formula 2 may be used in cases as small as ½ inch in diameter.

Before ramming, these mixtures should be moistened with a solution of one part of shellac in sixteen parts of alcohol. One part of this solution should be used to every thirty-six parts of composition. As the latter mixture is somewhat difficult to ignite, it is necessary to scoop out a little from the top of the torch and replace it with starting fire as shown in Fig. 17.

Starting fire

FIGURE 17

Torch with Starting Fire

GOLD AND ALUMINUM TORCH

Sodium perchlorate	13
Mixed aluminum	6
Dextrin	1

An aluminum torch of unusual brilliance and illumination, in the 1 inch diameter size and said to have 50,000 candle-power is made as follows:

Barium nitrate	38
Mixed aluminum	9
Sulfur	2
Petrolatum	1

Melt the petrolatum and mix with the barium thoroughly. Mix sulfur and aluminum separately; then mix this with the barium and petrolatum. Sift and mix twice. A starting fire for this is also necessary. The following formula for starting fire is suitable for the last three torches:

STARTING FIRE

Barium nitrate	4
Saltpeter	3
Sulfur	1
Shellac	1

Moisten slightly with alcohol before using.

The last three torches described should be allowed to dry for a week before being used.

Port Fires

These are small torches ⅜ inch in diameter and 12 inches long, used in exhibitions for lighting other pieces of fireworks. They are rammed with rod and funnel and two good mixtures are:

	Formula 1	*Formula 2*
Meal powder	1	
Sulfur	2	4
Saltpeter	5	5
Charcoal		1

Ship Lights and Signals

Bengolas

The bengola or blue ship light is a form of torch used mostly by ships in signalling for pilots. They consist of a strong water-proofed paper case 1½ inches in diameter and 4 inches long. There are 3 inches of composition and ⅓ inch of clay at the bottom; the balance is the socket into which the handle is fitted (Fig. 18).

FIGURE 18

Bengola

They should be rammed quite hard. The nosing should be of good, strong paper secured around the match with twine and the match should be piped where it passes through the nosing. The finished light may be painted with melted paraffin so as to protect it against the dampness of sea air. The following is an average formula:

Saltpeter	12
Sulfur	2
Antimony sulfide	1

Distress Signals

Distress signals are the same as the above except that they
burn red. The regular life-boat equipment consists of six or
twelve torches enclosed in a watertight copper can. The following formulas are suitable:

	Formula 1	*Formula 2*
Potassium chlorate	5	6
Strontium carbonate	1½	2
Shellac	1	1
Dextrin	½	

These should have a self-lighting end as for fusees.

Airplane Flare

This is a very powerful light of approximately 350,000
candle-power dropped from a moving plane at a height of about
5,000 feet. It is intended for illuminating a large area below
so as to permit photographing, bombing, reconnoitering, etc.

The light is suspended from a parachute so as to cause it to
remain in the air 5 minutes or more. It may be ignited either
by being lighted before being released or by the use of a small
windmill in its lower end, which operates by air pressure against
its vanes as it falls through the air. The stem of the windmill is
threaded and as it is turned by the wind its end eventually comes
into contact with a detonating cap, fires it and so lights a flare.
The sides of the flare are provided with wings which maintain it
in a perpendicular position while falling.

A typical flare is made as follows: A case is rolled of 4 sheets
of two-ply hardware paper, 24 inches long and 4 inches inside
diameter. When thoroughly dry, this is rammed by machine
with the following composition:

Barium nitrate	38
Aluminum	9
Sulfur	2
Petrolatum	1

Melt the petrolatum; mix with barium and sift. Mix aluminum and sulfur separately and sift. Mix both lots well and sift twice through an 18 mesh brass wire sieve.

When rammed, ream out ½ inch composition from bottom end of light and bore 4 holes, ¾ inch in diameter and ¾ inch deep into it. Ram this space with the following starting fire:

Barium nitrate	4
Saltpeter	3
Sulfur	1
Shellac	1

Dampen with alcohol.

When it is dry, prime well and lay two pieces of match across it; over this, place a cardboard wad with a 1½ inch hole through the center and secure it with muslin drumhead having a hole in the center in the same place, so as to expose the match.

The top of the case consists of a galvanized iron cover which may be riveted to the case before ramming. In its center is a ring for attaching the parachute cable. When assembling the light a small muslin bag is provided containing 70 grains 5F grain powder. To this is attached about 25 inches of piped quickmatch. Before the parachute container is attached, a small hole is made through it, 2 inches from the bottom and the match is passed through it. The container is glued into place and the match is led down the side of the flare to a small hole bored through it into the starting fire where the end of the match is inserted and fastened with a small tack. This is carefully covered with a strip of pasted paper. The other end of the match is led to the detonator.

The cable for suspending the light from a parachute consists of 36 inches of 6 strand wire. This is securely attached to the ring of the flare top before attaching the parachute container. Just over the joint it passes through a felt disk backed by tin on each side and of a size to fit the parachute container snugly. The cable is now carefully coiled on top of the disk and covered with about an inch of cotton hulls or bran, leaving about a foot of cable projecting to which the parachute is attached.

The parachute is made of silk and is about 15 feet in diameter when open. It is folded and inserted into the container which is lightly capped with a tin cover similar to a can top. Four fins of tin are attached to the sides of the parachute container. They are equally spaced and preferably riveted to the sides before the container is attached.[4]

Wing Tip Flares

These are used for signalling and to assist planes in landing. They are attached to the extremities of the wings and are fired by du Pont Flash Igniters, electrically controlled by the pilot.

They are made of a paper case with a wooden plug. The case is roughly 1½ inches in inside diameter and 4 inches long and rolled so as to be slightly tapered. This enables them to be securely inserted into holders on the plane wings.

Three-eighths inch at the front end and 1 inch at the rear end are left without composition when ramming, so as to provide room in front for the igniter and in the rear for the wooden plug.

Three-eighths inch of the front end is rammed with starting fire dampened with alcohol. This is primed and in the center of it the igniter is laid with wire leads carried around the outside of the case, to holes bored through it and through the wooden plug, emerging on back of the plug. A disc of cardboard is placed over the igniter and an outside wrapper of pasted paper is rolled around the entire assembly several times to secure it. When dry, it is ready for use.

Wing tip flares are generally made in white, red and green

---

[4] Faber, *Military Pyrotechnics,* **Vol. 2,** Washington, D. C., Government Printing Office, 1919.

using any of the formulas given for torches.

Rocket Smoke Tracers
(For Daylight Use)

Skyrockets are sometimes employed for conveying orders, information, maps, etc., to troops not easily reached otherwise. When used in the daytime, a smoke tracer is attached to the side of the rocket to assist in locating its course and destination. What is known as a 2 pound rocket is suitable, although larger sizes may be used.

The tracer consists of a paper case 4 inches long, ⅝ inch in inside diameter, rammed with any of the smoke formulas given in the section on smoke screens with a solid plug of clay at each end. Three or four holes ⅛ inch in diameter and about 1 inch apart are bored to a depth of ½ inch into the composition, through the side of the case. A strip of quickmatch is forced into these holes, primed over openings and a length of about 4 inches extending beyond the last hole is covered with match piping. The body of the case is now covered with a turn of pasted Kraft paper.

The tracer is fastened to the side of the rocket case with glue and wire on the side opposite to the stick socket and the piped section of match drawn around and inserted into the bottom opening of the rocket.

The best mixture for tracers is the following:

Saltpeter	5
Sulfur	2
Red arsenic	3

Very Pistol

The Very pistol is a variation of the Roman candle, discharging a star of various colors, as a signal to detachments of armed forces not otherwise within reach.

Toy Blue Lights

These are little lights ¼ inch in diameter and 6 inches long, made by rolling a light case as for lances. Cut the paper 2 x 6 inches, the 6 inch way running with the grain of the paper. Close one end as for lances. Bunch about two hundred into a bundle with string so that all the open ends are uppermost when the bundle stands on end. Now make up the following composition:

Saltpeter	5
Sulfur	2
Antimony sulfide	1

Mix and place on a large sheet of strong paper previously spread on a firm table. Set the bundle of blue light cases alongside of the composition on the paper, with the open ends up and pour a handful of composition on top of them. Shake the bundle so as to make the composition fall into the cases and repeat several times. Now, with both hands, raise the bundle of partly filled cases and bring it down on the table with a good blow. Repeat this operation of filling and pounding on the table until all are full. The ends may then be tucked in with an awl.

Roman Candles

These are probably the most popular type of fireworks made, from the sales point of view. Until some years ago, they were made entirely by hand, one at a time. Then a combination rammer making a dozen at a time was devised and finally the candle machine, which makes 6 dozen, was perfected. To make Roman candles by hand, roll the cases as described in the section on case rolling in Part II and have stars of different colors ready. Then make some candle composition as follows: [5]

[5] A. L. Due. Personal Communication.

Powdered saltpeter	18 lb.
Fine powdered charcoal	11 "
Flowers of sulfur	6 "
Dextrin	1 "
Water	1 gal.

After all the ingredients have been well mixed and sifted three times, add the water and mix again until the whole lot is dampened. Then force it through a 16 mesh sieve into cloth-bottomed trays and dry in the sun.

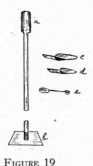

FIGURE 19

Hand Candle Rammer Outfit

Provide a ramming outfit as shown in Fig. 19 consisting of a rammer (a), a pin block (b), a composition scoop (c), a clay scoop (d) and a gunpowder scoop (e). The various parts must, of course, be proportioned to the size of the candle to be made. If you are making an 8 ball candle the pin of the pin block must be ⅜ inch in diameter. The rammer must be slightly smaller so that it can pass easily up and down in the candle case which also is ⅜ inch in diameter. The clay scoop should hold a level teaspoonful of clay, the composition scoop, a heaping dessertspoonful and the gunpowder scoop should be ¼ inch in diameter and ¼ inch deep. It may be made from a .22 caliber rifle shell, if desired.

Place an empty case on the pin; pour in a scoop of clay and ram it firmly with a light mallet. Remove the rammer and pour in a scoop of gunpowder. On top of this drop a star and, lastly,

a scoop of candle composition. Ram with about six light blows
of a small mallet. Remove rammer and pour in another scoop
of gunpowder, another star and another scoop of candle composition; ram as before and repeat this until the eight stars have
been loaded. The case should now be filled to within about 2
inches of the top. Remove candle and finish as described in the
section on *Finishing*.

Hand Combination Candle Rammer

This consists of the articles illustrated in Fig. 20: an iron pin
plate (e), iron funnel plate (f), a wooden guide board (d) and
three wooden shifting boards, i.e., clay board (c), star board
(b) and composition board (a). It also consists of a gun
powder box (Figs. 23 and 24) and a rammer (G, Fig. 21).
The construction of shifting boards can be readily understood
from detailed sketches (Fig. 21). The stop pins shown in upper
plates should be understood as being in lower plates; otherwise
slots would become clogged with composition while in use.

The holes in the upper board of the shifting boards are of a
size to hold just sufficient composition, clay, etc., for one charge
(a). This board slides a distance of about 3/4 inch, controlled by
pins (c–d). When the upper board is pushed back, the holes are
filled and just before the discharge it is drawn forward so that
the holes are in line with the holes in the lower fixed board (b).
The contents fall through the funnel into the candle to be rammed. The gunpowder box is described under *Candle Ramming
Machine*. It is smaller than that of the large machine and
of a size to correspond to the pin plate, etc. Finally, there is the
rammer (G, Fig. 21) consisting of eight steel rods with com-
pression springs fitted through a wooden handle bar, as shown
in H, Fig. 21.

This apparatus is used for ramming one to four ball candles
and can also be used for serpents and saucissons.

Place pin plate on a solid wood or concrete base; place guide
board over pins so that the holes more or less encircle the pins;

slip a candle case on each pin; place the funnel plate on top of the assembly and raise guide board so as to make cases center nicely and enter the funnel plate.

Now fill the clay board, composition board and star board. Place the clay board over the funnel plate so that the holes are

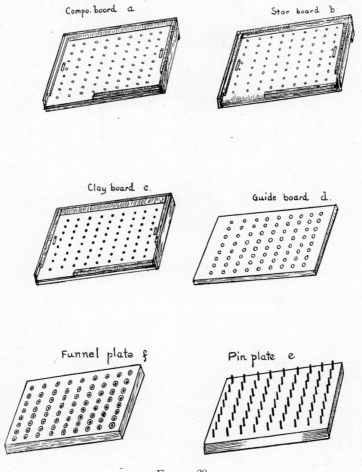

FIGURE 20

Hand Combination Candle Rammer Apparatus

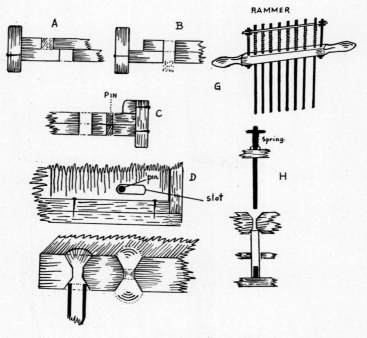

FIGURE 21

Detail of Shifting Boards

in line and shift, tapping lightly so that all of the clay falls through the funnel plate into the candles. With the rammer, give ten to fifteen firm blows through the row of holes. Put on gunpowder box and draw plate until a charge enters candles. Then take star board and shift. See that all stars have entered candles and then put on composition board. When this has been discharged, give about eight to ten firm blows with rammer, not quite as hard as for the clay. Now repeat the operation (if more than 1 ball candles are being made) until the desired number of stars have been used.

Candle Ramming Machine

The ramming machine illustrated in Fig. 22 is used principally for ramming Roman candles with 6 to 30 balls, but up to 3 ounce rockets may be rammed solid with it. The hollow center of the rocket is made by driving a spindle into it afterwards and will be explained later. Flower pots may also be rammed with this machine and the writer has adapted it to making 3 inch cannon crackers at the rate of seventy-two at a time. However, several sets of rammers of different lengths and thicknesses are required.

The frame is of cast iron about 7 feet high; the upright sides are $1\frac{1}{4}$ inches thick with V edges on the inner sides upon which the head block slides. The rammer assembly is fastened to the head block by stud bolts. The guide board (C) is made of $\frac{1}{2}$ inch lumber and serves to keep the rammers properly in line. This board is loose enough to slip up and down on the rammers while the machine is in use. The pin plate (D) rests on the base of the machine and slides into place from in front and is retained by short lugs in the rear. Several of these plates, corresponding to the rammer assemblies mentioned above, are also required. The pawl (E) holds the rammers up while the articles to be rammed are arranged below. When all is in place and the first charge of clay (in the case of Roman candles) is placed into the cases, someone pulls the rope attached to the head block, which serves to disengage the pawl. The rammers are now lowered slowly until they enter the funnel plate. The rope is released and, as the rammer head falls, it rams the clay in the bottom of the cases. From five to fifteen blows are usually required to ram each charge.

If the composition becomes so dry that it will not pack firmly it should be dampened with a very little water. The stars should be hard and dry and free from star dust which can be sifted out by shaking the stars in a coarse sieve. The floor of the ramming

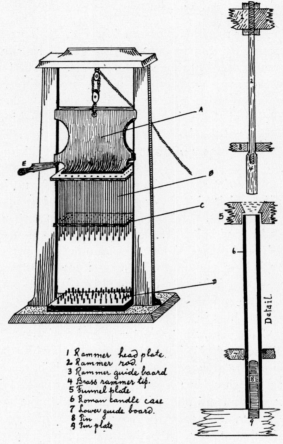

1 Rammer head plate.
2 Rammer rod.
3 Rammer guide board
4 Brass rammer tip.
5 Funnel plate
6 Roman candle case
7 Lower guide board.
8 Pin
9 Pin plate

FIGURE 22

Ramming Machine

room should be kept scrupulously clean and free from all accumulated composition, etc., to guard against accidents from friction of shoes or otherwise. A rubber floor-covering is desirable. Remove all rammed articles frequently so as not to accumulate too much material.

It should be noted here that when cutting the paper for machine-rammed Roman candles a thin V-shaped slip should be cut from one end (a, Fig. 22 a) of the sheet at the side which is nearest to the operator during the rolling process. The object of this is to form a somewhat funnel-shaped end to the case which will greatly assist in easy ramming. This end, of course, must be uppermost when the case is in the machine (b, Fig. 22 a).

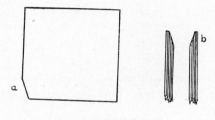

FIGURE 22a

a. Paper, Cut for Ramming b. Same, When Case Is Rolled

The funnel plate is made of cast iron 1 inch thick; the other dimensions are the same as the head of rammer assembly. It is drilled with 72 holes in six rows of twelve each, corresponding with the inside diameter of the candle to be rammed, and spaced the same as the rods in the rammer head. These holes are countersunk on the upper side of the plate to a depth of one-third of its thickness so as to give them the shape of a funnel. The under side is counter bored, somewhat larger than the outside of the candles which, before being rammed, slip into these recesses and thus are held in place while the machine is being operated. The funnel plate is supported in the ramming machine by an adjustable frame attached to the sides of the machine. This permits it to be moved up and down as required to fit the various lengths of cases to be rammed. This frame is not shown in the drawing of the machine.

Powder Box

The powder box (Figs. 23 and 24) is made of brass $\frac{3}{16}$ inch thick. Its construction will be readily understood from the sketch. The bottom consists of three plates, each $\frac{1}{8}$ inch thick, drilled with $\frac{3}{16}$ inch holes spaced the same as those in the funnel plate. The holes in the upper and middle plates are $\frac{1}{2}$ inch nearer the rear of the box than the holes in the bottom plate. The upper and lower plates are fixed but the middle plate moves forward and backward $\frac{1}{2}$ inch. When it is pushed back the holes in it are in line with those in the top plate so that, when the box is filled with rifle powder, the holes in the middle plate become filled. When it is drawn forward later the holes come in line with those in the bottom plate and the little powder charge in each hole falls out into the candle below it.

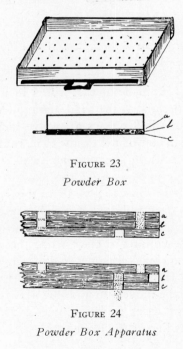

FIGURE 23

Powder Box

FIGURE 24

Powder Box Apparatus

To facilitate the use of this box, it is placed on the adjustable stand (Fig. 25), on which it can be raised to the desired height for the work on hand. This stand is made of light lumber and preferably on rollers so that it may be moved into position and out of the way, as desired, between charges.

The shifting boards of a big machine follow the same principle as those of the hand ramming machine, in regard to arrangement of holes, etc. Boards of different thicknesses must be provided to hold the required amounts of composition for

FIGURE 25

Adjustable Stand

the different sizes of candles rammed. The holes in star boards should be slightly larger than the stars so that they fall through easily. The stars for Roman candles should also be somewhat longer than their diameter as this makes it easier for them to fall into place when filling the shifting board.

To fill these boards, a scoop of clay or composition or a handful of stars is thrown on top; the board is shaken until the holes are evenly filled and the surplus is allowed to slide off into the composition or other trays. The extra pin plates and boards are filled as quickly as needed, and the loaded candles removed. A very large number can be charged in a day by one machine.

A pin plate of candle cases is slipped onto the base of the machine; the funnel plate is lowered on top of it; the guide board is raised, causing the ends of the cases to enter the funnel plate which is fastened in place by set screws or thumb bolts on the sides of the frame. Permit the rammer head to descend sufficiently to see that all is clear. It is drawn back up into place and a shifting board of clay slipped over. Its contents are discharged into the candles, a slight jar being given to empty all the holes. The rammer head is now dropped some ten to twelve times to set the clay and is then withdrawn to its original position. The powder box is slid across the funnel board and, as the result of a pull on the handle of the center plate, a charge of gunpowder enters the candles. After removing the powder box, a board of stars is shifted into the funnel plate.

Care must be taken to see that all stars have slipped through the funnel plate into the candles. Next, a board of composition is discharged the same way and all of it is rammed with about eight to ten blows. This operation is repeated as often as the size of the candle requires. When the last charge of composition has been rammed, the pin plate of candles is removed, unloaded and refilled with empty cases. Another pin plate of empty cases is slipped into place in the machine and filled while the candles are being removed from the first plate.

Batteries

A very effective piece of fireworks (Fig. 26) is easily made by taking a wooden box about 2 inches longer than the candles

FIGURE 26

Battery

to be used and filling it with about 3 dozen 8 to 10 ball Roman candles.

The space above the candles in the box is filled with scraps of match. One piece is allowed to hang over the side and a piece of cardboard protects the top until it is ready for use. The cardboard must be removed before lighting.

Bombette Fountains

Bombette fountains are an effective combination of candles and floral shells packed in a box as shown in Fig. 27. All candles are lighted by scraps of match in the top, but the floral shells are matched, as shown in Fig. 27, so as to fire alternately one at a time, during the burning of the candles.

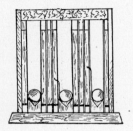

FIGURE 27

Bombette Fountains

Union Battery

Another interesting use for Roman candles is in the so-called Union battery (Fig. 28). No doubt, this originally consisted of

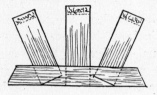

FIGURE 28

Union Battery

a battery of red, white and blue stars. It is now used effectively with candles of various colors.

Bengola Battery

Fireworks displays are often started with a row of vari-colored lights or bengolas placed about 25 feet apart in front of the set pieces. When these are supported by a fan of candles or gerbes a very effective display is produced. The bengolas are lighted first and when they are half burned the candles or gerbes are lighted (Fig. 29).

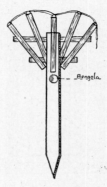

FIGURE 29
Bengola Battery

Sky Rockets

Next to Roman candles these are perhaps the most popular articles of the pyrotechnical craft and, on good authority, apparently antedate the candle. So much has been written about sky rockets that any detailed description would be superfluous. The French, particularly, have left a most complete history, sometimes amusing, in view of the present status of rocket manufacture. The rocket consists of a tube of paper rammed with suitable composition, its lower end choked to about one-third of the diameter of its bore and having a hollow center extending upward through the composition to about ¾ inch of the top. A stick attached to the tube serves to balance it while ascending. Roughly, the composition of a rocket, that is, the portion of it that is burning while it is ascending, should be seven times its diameter in length. Six-sevenths is pierced through the center while one-seventh is solid and acts as a fuse to communicate the fire to the heading when the rocket reaches the highest point of its flight.

The tube is made of strong paper, preferably 3 turns of hardware paper on the inside with 4 or more turns of straw board or Kraft paper on the outside. A good rocket case can also be made of heavy rag or building paper, if it is properly rolled with good paste. The process of choking the case and ramming in a mold has been practically discontinued. An average model for a 1 pound rocket is given in Fig. 30 with a corresponding set of ramming tools in Fig. 31. The spindle is one-half of the actual size, whereas the ramming tools are one-third of the actual size.

Good rockets should be uniform, all those of one caliber ascending to the same height and bursting at about the same time. This is particularly desirable in bouquets of 100 or more, fired simultaneously, or a straggling effect is produced.

Most rockets larger than 3 ounce rockets are rammed singly or by gang rammers as shown in Fig. 32. Hydraulic rammers are also in use.[6]

The gang rammer is quite efficient and with it one man can make a large number of rockets in a day. *A* shows the spindle block; *B* is the guide board for inserting the ends of cases into the funnel piece *C*. *D* shows a set of rammers while *E* is the set of scoops for charging all of the six cases at once. The latter is easily made by cutting brass shotgun shells and soldering them

[6] Faber, *Military Pyrotechnics,* **Vol. 2,** p. 39, Washington, D. C. Government Printing Office, 1919.

T. L. Davis, *Chemistry of Powder and Explosives,* p. 77, New York, John Wiley and Sons, 1943.

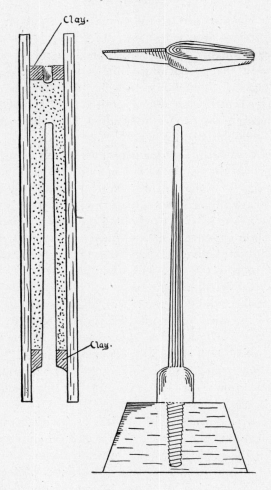

FIGURE 30

One-pound Rocket Section Model and Spindle

to a brass rod as indicated. Details of the funnel piece and hollow pin rammer used in setting top clay charge are shown, (*F* and *G*, Fig. 32).

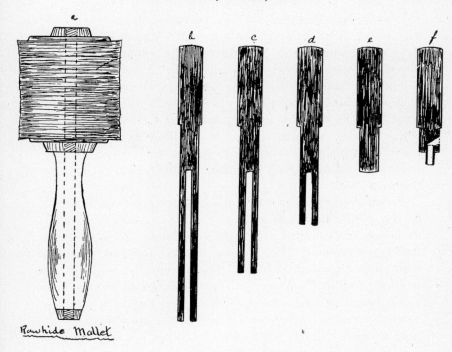

FIGURE 31

One-pound Rocket Ramming Tools

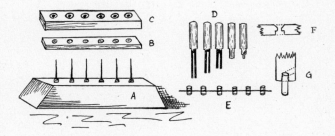

FIGURE 32

Gang Rammers

One- to three-ounce rockets are rammed solid on the candle machine or by hand and the hollow center is made by driving a steel spindle into them afterwards. These must have their lower ends choked as explained previously. An efficient way of making the hollow center is to use a mortising machine and replace the chisel with a spindle as mentioned above. A V-shaped block is set on the table of the machine in such a position that when a rocket is placed against it, it will be in just the right place for the spindle to enter it. A step on the pedal of the mortising machine will force the spindle into the rocket, forming the hollow center.

To ram rockets from 4 to 8 ounces singly, the case is slipped on the spindle illustrated under *sky rockets*. A scoop of clay is shaken in and rammed with eight good blows of the mallet on the longest rammer. Then a scoopful of composition is rammed with about eight lighter blows. This is repeated until the case is filled to about 1 inch from the top. Shift rammers as it becomes necessary to use shorter ones. There should be 1 inch of solid composition above the top end of the spindle. Now the final charge of clay is put in and the hollow pin rammer used. This sets the clay while leaving an opening for the fire to reach the heading. Care must be used to see that the hollow tube just pierces the clay. If it does not go through, the heading will fail to fire; if it goes through too far, the heading will fire prematurely or the rocket may blow through before rising.

The following are good compositions for rockets of the different sizes given:

	1 to 3 Ounces	4 to 8 Ounces	1 to 3 Pounds	4 to 8 Pounds
Saltpeter	18	16	16	18
Mixed coal	10	9	12	12
Sulfur	3	4	3	3

If rockets burst before ascending add more coal; if they ascend too slowly add more saltpeter. For the smaller sizes use fine coal; for larger ones use coarser coal in proportion to the diameter. In 4 to 8 pound rockets use partly granulated salt-peter.

All rockets larger than 3 ounces are provided with a cone containing the heading. These are made as follows.

Sky Rocket Cones

Turn out a *cone form* on the lathe shaped as shown in Fig. 33.

FIGURE 33

Sky Rocket Cone Former

Cut some stiff paper to the shape of one-third of a circle, the radius of which should be 3 inches for a 1 pound rocket. Lay it on the table before you with the round side toward the right. Paste the straight edge farthest from you and place the form on it with the point toward the left and about ⅜ inch from the point where the two straight edges meet (the apex of the triangle). Now roll the paper around the cone form commencing with the unpasted edge. When it is finished slip it off and dry.

Heading Rockets

Prepare a board with holes through it about 1¼ inches in diameter and raised from the table about 3 inches as shown in Fig. 34.

Place cones in these holes with the point downward. Fill them

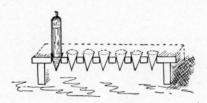

FIGURE 34

Board for Heading Rockets

about half full of stars, gold rain, etc., a little meal powder and charcoal or candle composition. Apply gum to the upper edge of a rocket and stick it into one of the cones. Raise carefully out of the hole and press the cone evenly into place. Set aside to dry. The rocket may now be wired to the stick and is then ready for use. In the case of shelf goods, the rockets are, of course, papered and matched before attaching cones.

Short Stick Rockets

These are the same as long stick rockets except that a stick only one-third the regular length is used. This makes a less cumbersome rocket which ascends higher. However, a wing tab must be attached to the stick as shown in Fig. 35. Cut a piece of cardboard about 3 inches long, 1½ inches wide at one end and ½ inch at the other. Smear a little dextrin on one end of the stick; place the tab on it, wide end down, and drive a 2 ounce tack through it in the middle.

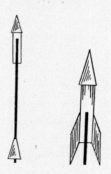

FIGURE 35

Stick with Wing Tab and Stickless Rocket
(Predecessor of the V-2 of today.)

When dry it is ready for use. These rockets are much easier to carry about but require more care in firing to get them started straight. The one illustrated at the right has no stick at all, only four wings, upon the ends of which it rests when lighted.

When the bottoms of brass rocket rammers become worn

from use they may be reconditioned by battering them until they are again full size on the end.

There is a great variety of so-called fancy rockets, in which the heading is not confined to a single burst of stars, etc., but is supplemented by many other beautiful and interesting effects, some of which will be described here.

Rockets Without Sticks

The present-day *bazooka* is the adaptation of the winged rocket which was devised many years ago. Its construction and operation will be readily understood from the accompanying sketches.

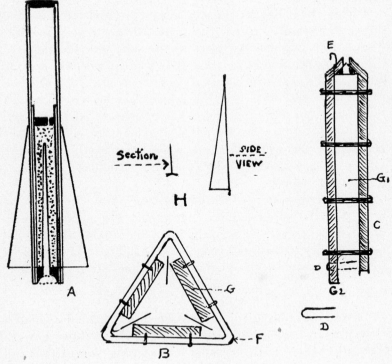

FIGURE 35a
Apparatus for Rockets without Sticks

Rockets for this purpose should not be smaller than 1 inch in internal diameter (3 pounds). The heading, if any, must be inserted in the top of the rocket which must be of the same outside diameter for its entire length. This may be achieved by making the rocket in the regular way, as previously described, and then rolling around it a jacket of about 3 turns of heavy paper, approximately twice the length of the rocket, flush with its lower end. This will leave a receptacle at the top for the parachute or other desired heading (A, Fig. 35a). To the lower half there are attached three wings spaced at equal distances of 120° around it (A). These may be made of tin or stiff paper and in either case may be fastened by strips of glued or pasted paper (H).

To fire the rocket it will be necessary to make a launching device as follows: Procure two pieces of board ½ x 2 inches D4S 6 feet long and one piece of the same size but 6 feet 3 inches long (G_1 and G_2). Make four triangular bands of strap iron ½ inch wide, ⅛ inch thick and 12 inches long (B). On each of the three flat sides (F) drill 2 holes to take screws. Insert the boards, one at a time, and fasten in place as in sketch (B). This leaves three slots inside of the launcher, between the boards, for the wings to move when the rocket is launched. On top of one board attach a hook (E) for hanging the launcher to a suitable support.

Through the bottom of the long board (which should project at the bottom) bore two holes (D) to take the supporting fork (D). When everything is ready, insert the rocket, allowing the bottom to rest on the fork. It may now be lighted and will ascend to a much greater height than a rocket with a stick.

Willow Tree Rockets

These always beautiful rockets are simply made by filling a large rocket head with pieces of Japanese star and a weak bursting charge. If the bursting is too strong, some pieces will fail to take fire. In order to permit carrying a large charge of the star, which is very light, the heads are usually made bulging toward the cone.

Prize Cometic
(Shooting Star Rockets)

To make shooting star rockets place four or five 4 ounce rockets, without sticks, in the head of a 6 pound rocket along with a handful of box stars. A few #1 stars are also placed in the top of each 4 ounce rocket with a pinch of grain powder, after which they are securely capped without cones.

Golden Cloud Rockets

For these, each rocket head is filled half full of gold rain and aluminum stars. The weight of a rocket head must be proportional to the size of the rocket. Obviously, a heading of heavy stars must be smaller than one of lighter materials.

Boom Rockets

These have one or more small maroons in the head in addition to a few stars.

Electric Shower Rockets

These are made by filling a small head with electric spreader or granite stars. Owing to the weight of this heading only a small quantity can be used.

Bombshell Rockets

These have a small shell with a very short fuse attached to the top of the rocket. A few stars are in the head itself. These burn before the shell bursts.

Whistling Rockets
(Calliope Rockets)

The heads of these rockets are filled with whistles, made as described under that caption. In addition, a few colored stars are added.

Liquid Fire Rockets

These are among the most beautiful pyrotechnical effects known to the art. Take a 3 pound rocket and fill the space above the clay with gunpowder. Glue onto this a circular piece of perforated paper. Waxed or cellophane paper may also be used. Roll on a head of about 3 turns of strong Manila paper, pasted only on the edge and about 6 inches long. Cut some sticks of yellow phosphorus under water with a chisel into pieces about $\frac{3}{4}$ inch long. Get some $\frac{1}{4}$ pound tin cans, punch a number of holes into the bottom of them and fill them with the pieces of phosphorus, conducting the entire operation under water, preferably in a wooden bucket.

When ready to fire the rockets, remove one of the cans from the water, allow to drain for a few seconds, empty the contents into one of the rocket heads, tuck in the top of head, place in the rocket stand and fire *at once*. Great caution must be observed because of the inflammable nature of phosphorus. There is absolutely no danger if the above directions are followed. A similar effect may be obtained by the use of metallic sodium. In this case the work must be done under kerosene oil instead of water.

Parachute Rockets

To launch successfully a parachute from a shell or rocket requires the greatest care and skill and patient attention to every detail. The light fabric may fail to unfold or be torn or burned in its exit from the tube in which it is placed. To begin, cut some very light Japanese tissue paper into squares of 18 inches and rub them thoroughly with powdered soapstone. Cut four pieces of strong linen twine or shoemakers' thread about 18 inches long. Twist the corners of the tissue squares a little and tie a thread to each. Tie the other four ends of the threads in a knot.

The parachute is now ready to fold. In one hand take the

LIQUID FIRE ROCKET

knot where the four strings meet and in the other hand take the top of the parachute by the center. Draw the hands apart until the paper folds itself together and lay it on a table in front of you. Straighten out the four folds, two each way, and fold them again laterally toward the center about five or six times like the bellows of an accordion until the pile is about 1 inch wide. Now roll this up lightly, beginning at the small end or top, until you come to the strings. Then wind the bunch of four strings lightly around the bundled parachute until it just fits the head of the rocket for which it is intended.

For making the light or flare, ram a short case $3/4$ inch in diameter and 1 inch long with box star composition, first putting in a small charge of clay, and prime the other end. Over the clayed end glue a cardboard disk slightly smaller than the inside diameter of the rocket head, having first passed a wire through the case under the disk so as to form a loop on top as shown in Fig. 35b. Pass about 18 inches of stout linen twine or sized cord through the wire loop and tie the other end to the knot on the parachute where the four strings come together.

Roll a piece of naked quickmatch, about 18 inches long, into a ball and place it in the bottom of the rocket head for a blowing charge. On top of this drop the primed end of the parachute light and over it place a small wad of cotton waste and over this a small quantity of cotton hulls or bran. Now slip in the

FIGURE 35b

Flares

parachute around which the strings have been lightly wound. Fill the space around the parachute with bran and secure the top of the rocket head very lightly so that the parachute will be thrown out with the least possible effort when the rocket is discharged.

Chain Rockets
(*Caterpillars*)

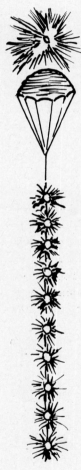

FIGURE 35c

Chain Rockets

If you have succeeded with parachute rockets you may now attempt this modification of them which is more difficult. Their beauty compensates for the trouble required in their preparation. A parachute several times as large as the one first described is made in practically the same manner but preferably octagonal with the separate pieces sewed together.

Instead of one light, a dozen or so of differently colored lights are attached to it. This is called the chain and to launch it successfully from either rocket or shell is about as difficult a task as the pyrotechnist is called upon to perform. For the lights composing the chain, ordinary lances may be used. A 4 pound rocket requires twelve lights. Procure a large strong waxed cotton string about 18 feet long and attach to it the lances at intervals of 18 inches by taking two half hitches around the bottom ends. It is best to make special lances for this purpose filling the first $\frac{3}{8}$ inch with clay.

When all are fastened, tie one end of the chain to the parachute and at the other end begin to wind up the slack between the lances. Wind each lance with the slack between it and the next one to it, winding as smoothly as possible without lapping the twine anywhere. As each one is wound lay it against the one before it until the twelve are in a round bundle. Then wind a few turns around the entire bunch on the upper end to hold it together. At the bottom end of the bunch take 2 turns of light cord not more than $\frac{1}{4}$ inch from the end. This is to hold the lot together until all the lights take fire when this cord burns off and the chain unwinds in the air. A cardboard wad, fitting easily in the rocket head and with a hole punched through its center, is placed on top of the bunch of lances and a piece of match passed through the hole in it so as to touch them. This may be fastened in place with one or two small tacks.

Now prepare the rocket head for the reception of the chain, as directed for parachute rockets, by placing 2 feet of naked match (without piping) in the bottom of the head for the blowing charge. Slip the bunch of lances in on top of this, with another paper disk over it, through which the line runs to the parachute. Put in a large wad of cotton waste, then carefully fold the parachute, as described above, and pack with bran. Cap the rocket as lightly as possible and if all the directions have been carefully followed, the chain will most likely come out successfully. A few trials, however, are generally necessary. Sometimes four light sticks are inserted in the rocket head, alongside of the parachute, equally spaced, with the lower ends resting on the strong wad under the bunch of lances and the other ends against top disk over the parachute. This is to keep the parachute from being injured while being expelled by the blowing charge.

Bouquet or Flight of Rockets

These are made by firing one hundred or more rockets, all at once, from a specially prepared box (Fig. 36).

Take three boards of ½ inch lumber, 12 inches wide and 4 feet long; clamp two of them together and with a ½ inch bit,

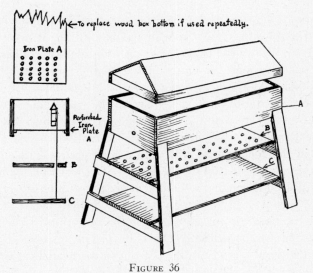

FIGURE 36

Box for Rocket Flights

bore five rows of 20 holes each 2 inches apart and beginning
2 inches from the sides and ends. This will make 100 holes
through the boards (A and B, Fig. 36). Now construct a box,
the bottom of which is made of one of the boards with holes
through it as shown in (A). Attach four legs to a box, about
4½ feet long. At 1½ feet from the bottom secure the other
board (B) with the holes in it so that a rocket passed through
a hole in the box bottom may be steadied by passing through the
corresponding hole in the lower one. Fit the third board (C)
in the legs also, about 6 inches from the ground to make a rest-
ing-place for the rocket sticks and so as to hold them about 1
inch above the bottom board (A) of the box. This is to permit
the fire to reach all the rockets instantly when lighted.

Flight rockets used this way need not be matched, only
primed. A little loose gunpowder thrown on the bottom of the
inside of the box and a piece of match passed through a hole to
the inside to fire it are all that is required. If a top, covered
with canvas is fitted to the flight box, it may safely be left in
case of rain until required. Some pyrotechnists make flights by
stringing rockets in a row on slats provided with nails to hold
them apart, but the effect is not as good.

Rocket Stand

The best method of firing sky rockets is from a wooden
trough constructed of two light boards ½ inch thick, 4 inches

FIGURE 37
Rocket Stand

wide and 6 feet long (Fig. 37). These are nailed together so as
to form a gutter and are supported by two legs. If the boards
and legs are hinged as shown, the trough may be folded and
easily carried about.

Rockets in Warfare
(Rocket Bombs)

The deadly contrivance sketched (Fig. 37a), developed during World War II, really has no place in a book devoted to the
production of pleasure-giving devices. However, as it is based
on the principle of the pyrotechnical rocket and is of such great
interest, an outline of its construction and operation is given
here as a matter of record.

Some of these jet-propelled missiles are so small that they
are fired from a launcher only a few inches long; others found
in Germany were 40 feet high. The smaller ones may be
launched from a plane and are more accurate and penetrating
than a bomb. One of 5 inch diameter can pierce 4 inches of
steel. The larger ones carry up to 1200 pounds of explosives.
Some are also radio controlled.

The largest bombs consist of a cylinder built of steel, magnesium and aluminum, with three stabilizing fins attached at
equal distances around the lower portion. They are believed
to reach a height of several hundred miles from the earth.
It was claimed that they could be fired from Europe to America.

The sketch gives a general idea of their construction. The
driving force consists of gasoline or alcohol and oxygen.
The exact construction has not yet been revealed by the War
Department but some captured specimens may be seen in the
Museum of Industry in Rockefeller Center. They are erected
on the ground in a perpendicular position and ignited electrically.

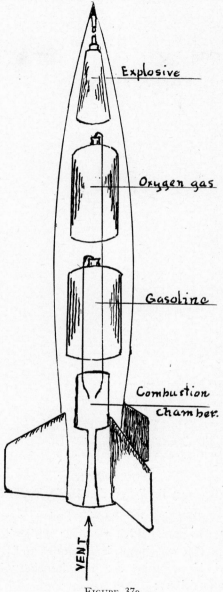

Explosive

Oxygen gas

Gasoline

Combustion chamber.

VENT

Figure 37a
Rocket Bomb

FIGURE 37b

V-2 Soars in Flight Test

A 46 foot, 14 ton remade German V-2 rocket just after it was fired in a flight test at White Sands Proving Grounds, N. M. May 10. It went an estimated 75 miles above the earth and landed in the New Mexico desert 39 miles from its launching platform.

Reprinted by courtesy of Press Association, Inc.

Tourbillions

(Geysers, Whirlwinds, Table Rockets)

This is a modification of the sky rocket, somewhat on the order of a gyrocopter, which revolves while it ascends to a height of about 100 feet, in a spiral manner and without a stick. They are made by ramming a 3 pound rocket with one of the following compositions:

	Formula 1	*Formula 2*
Saltpeter	8	5
Meal powder	7	12
Charcoal	2	3
Sulfur	2	3
Steel filings	3	

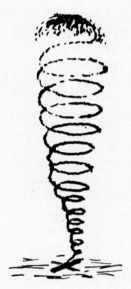

FIGURE 37c

Tourbillion

Both ends of the case are stopped tightly with clay. Four holes, ¼ inch in diameter, are bored into the side of it. Two holes are bored in the bottom, 3 inches apart or 1½ inches each way from the center, and one hole on each side, opposite to one another as shown in Fig. 38. A piece of curved stick, as

FIGURE 38

Position of Holes for Tourbillion

long as the tourbillion, is nailed to the bottom of the case, the concave side against the case, exactly in the center and at right angles with it, using a round-head nail as long as possible without letting it come through the case on the opposite side. The holes are primed and while still wet they are matched by tacking a piece of quickmatch to one of the bottom holes, passing it to the nearest end hole; then over the top to the other end hole and finally to the other bottom hole. A small hole is now made in the match pipe where it passes over the top of the case, just in the center, into which a short piece of match is slipped for lighting. If good sharp match is available, it may be used unpiped for matching tourbillions and covered by a strip of *Foudrinier* paper pasted overall.

To be fired, a tourbillion should be laid on a wide board or smooth surface, stick down, and lighted with a long portfire. Small tourbillions are sometimes made by boring only two holes in the under side of the case, at an angle of 45° *from* the bottom, on opposite sides, but those with four holes, especially in the larger sizes, are safer and more likely to function.

Large tourbillions are sometimes beautified still further by placing a few stars in the ends of the case, outside of the clay, boring a small hole through it, adding a little gunpowder and securing the top with strong paper and a wad. When the tourbillion reaches the end of its flight, the stars are thrown out with fine effect.

Flying Pigeons

This amusing piece of fireworks is easily made in its simplest form by securing two rockets, with their openings pointing in opposite directions, to an empty case as shown in a, Fig. 39.

The rear end of one is connected by a piece of match to the front of the other. A piece of thin rope or telegraph wire is stretched between two posts about 20 feet high and 200 feet apart. One end of the wire is first slipped through the empty case forming the middle of the pigeon. On lighting the first

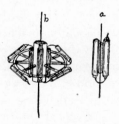

FIGURE 39

Flying Pigeons

rocket, the pigeon will run along the line until the other rocket lights, when it will return to its starting point.

A more effective form of pigeon (b) is made by constructing a frame as shown. This consists of a vertical wheel with a heavy slotted hub. A set of four wheel cases is fastened to the rim and four 1 pound rockets are secured to the long slots in the hub, two pointing each way. The pigeon starts with one of the wheel cases, the rear end of which is connected to one of the rockets. This in turn is matched to the second wheel case and that to the next rocket, pointing in the opposite direction, and so on to the last rocket.

English Crackers
(*Grasshoppers*)

Cut some good 20 pound, 24 x 36 inch Manila or Kraft paper into strips 4 inches wide and 12 inches long. If cut with the grain of the paper as it should be this will give eighteen cuts from one sheet. Roll them into short tubes as directed for match pipes, getting the opening at one end somewhat larger than that at the other. This may be done by rolling a V-shaped strip of paper on one end of a rod. When a considerable quantity of these tubes have been rolled, close the small end by twisting or folding it over. Dry them in the shade and put about 12 dozen in a bundle, all the open ends one way.

Stand the bundle on a large sheet of paper with the open ends up and pour FFF grain powder into all the tubes until they are full. Jolt the bundle occasionally so that none of the tubes is only partly filled. Then draw the tubes out, closing the top as you did the bottom and wrap them all in a wet towel, setting aside in a damp place for several hours. A good way is to take a long cloth, wet it well, spread the loaded pipes loosely on it and roll it up so that each pipe will touch a part of the wet cloth. The tubes should be thoroughly moistened, but not too wet, before proceeding further. When this condition (on which the whole success of the operation depends) has been reached run them through a clothes wringer or other roller so that they will be somewhat flattened. The exact amount of flattening can only be ascertained by experiment.

Now take a piece of wood, 1 inch thick, 4 inches wide and 18 inches long. Notch out a piece, as shown in Fig. 40, 1½ inches wide and 6 inches deep. Procure a dozen pieces of stiff wire 4 inches long. Lay the lower ends of a half dozen of the damp pipes across the bottom of the notched board which has been fastened in an upright position to a bench.

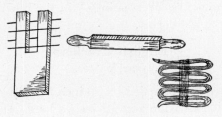

FIGURE 40

Apparatus for English Crackers

On top of these and against one side of the board lay a wire and bend the pipes over it until they now point in the opposite direction. Lay another wire as before but on the opposite side and repeat the operation until the entire length of pipes has been folded up. Then take a bar of wood, shaped as shown in the sketch and, holding one end in each hand, press the folded pipes down as hard as possible so as to have the turns well formed.

Now lift out the folded bunch, wires and all. Remove wires, fold bended pipes one by one, in the hand, and with shoemakers' linen thread secure them by wrapping a half dozen turns around the bunch and finally, a few turns between the folds.

Strip off one end so that the powder is exposed and prime it with a little wet powder or match it, or twist up the end with *touch paper* made by coating unglazed paper with a solution of saltpeter and drying. When dry, crackers are ready for use.

In this country, this item, usually called grasshoppers, is made by taking exhibition match with closely fitting pipes and cutting it into suitable lengths, of about 15 inches. A ½ inch of bare match is withdrawn from one end of the pipe while the other end is closed by twisting. A bundle of these is now wrapped in a wet towel as above and allowed to soften. When the pipes have become so damp that they will not tear when bent at right angles they may be folded as described. This method is much faster but does not make as good a cracker.

Pinwheels

FIGURE 41

Pinwheels

For making pinwheels proceed in the same way as for English crackers using the following compositions instead of gunpowder for charging:

	Formula 1	Formula 2	Formula 3	Formula 4	Formula 5
Meal powder	2		10	8	2
FFF Rifle powder		8	5	8	
Aluminum				3	
Saltpeter	12	14	4	16	1
Steel filings		6	6		
Sulfur	2	4	1	3	1
Charcoal	4	3	1	8	

When they are dampened and rolled, punch out a lot of round pieces of 60 pound strawboard or cardboard, with a hole through the exact center. Then get a piece of brass, the same size as the cardboard centers and fasten it to the work table. Lay one of the centers on this brass plate and, taking a filled pinwheel tube, press the smallest flat end against its edge. Twisting it around the disk with the right hand while the left hand feeds the tube as it is being wound on, continue until all of the tube is rolled around the center. The brass plate should be half as thick as the finished pinwheel so that the cardboard center will be held just in the center of the finished pinwheel, while it is being twisted.

Now have some boards prepared with strips of wood ¼ inch square nailed to them, the same distance apart as the width of a pinwheel when it is lying down. When the wet pinwheel is twisted up as described above, lift it off the brass plate and set it between two of these strips on the board so as to keep it from untwisting, and with a brush, put a drop of glue across the pipes and onto the center disk, at four equidistant points. When the pinwheels have dried they may be removed from the board and are ready for use.

Serpents, Nigger Chasers
(*Squibs*)

These are light strong cases, 3 to 5 inches long, crimped at one end and charged with a sharp composition which is strong

FIGURE 42

Serpent

enough to cause them to move around rapidly on the ground or in the air while burning. They may be made from #140 strawboard, or similar paper, and crimped while still wet. They may be rammed singly with rod and funnel or in batches of seventy-two at a time with the hand combination rammer. Alternate compositions are:

	Formula 1	Formula 2
Meal powder	3	3
Saltpeter	2	5
Sulfur	1	1
Mixed coal	1½	¾
FFF Grain powder	4	3

Saucissons

FIGURE 43

Saucisson

These are very similar to serpents but somewhat larger and always end with a report. Also, they are not used to *run* on the ground but are fired from mortars in bunches. The usual length is 3½ inches with a diameter of ⅜ to ½ inch, rolled and crimped like serpents, but with a heavier case. Ram with:

Meal powder	4
Saltpeter	2
Sulfur	1
Fine coal	1½

For exhibitions, about 3 dozen of these are put in a paper bag with 3 ounces of blowing charge composed half of meal powder and half of grain powder. A piece of match a yard long, bared for 1 or 2 inches, is stuck into the mouth of the bag and secured tightly with a string. When ready for use, it is inserted into a mortar and ignited.

For stock work, a paper mortar is made by rolling six or eight thicknesses of heavy strawboard, 12 inches wide, around a form $2\frac{1}{2}$ inches in diameter. A wooden bottom is fitted and a mine bag made as described under *mines*. The saucissons are placed in it with the blowing charge, around a 10 ball candle from which the bottom clay has been omitted. This is placed in the mortar with a daub of glue on the bottom of the bag. A top is fitted as for mines and when papered and striped it is ready for the market.

Mines

These are small paper guns from 1 to 3 inches in diameter, on the bottoms of which are placed small bags of stars, powder, etc. They are fired by a mine fuse or Roman candle in which the initial charge of clay is replaced by one of the candle compositions. The bottoms are turned out of wood. The tubes are made by tightly rolling six to twelve thicknesses of strawboard or other heavy paper around a suitable form.

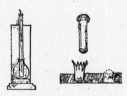

FIGURE 44

Apparatus for Mines

The approximate sizes are as follows:

No.	Height, Inches	Diameter Inches	No. of Strawboard #
1	4	$1\frac{1}{2}$	1 sheet 140
2	$4\frac{3}{4}$	$1\frac{3}{4}$	1 " 120
3	$5\frac{1}{2}$	2	1 " 100
4	7	$2\frac{1}{4}$	2 " 100
5	$8\frac{1}{2}$	$2\frac{3}{8}$	3 " 100
6	10	$2\frac{7}{8}$	4 " 100

Number 1 mines take a 1 ball mine fuse; #2 mines a 2 ball fuse, etc.

Mine bags are made by first boring a number of holes into a thick board (for #1 mines, 1¼ inches diameter and 1½ inches deep). Then make a punch with rounded edges (for #1 mines, 1 inch diameter); press a piece of strong paper (about 4 inches square) closely around the end and shove it into one of the holes in the board (Fig. 44). Remove punch; insert a mine fuse and around it put a half ounce of stars and a teaspoonful of blowing charge. Squeeze loose ends of the bag around the fuse and secure with a wire or piece of string. Now daub the bottom of the bag with a little glue or dextrin and insert it into the gun. A top is provided as follows:

Adjust an ordinary washer cutter to the requisite size so as to cut a piece of paper board with the outside diameter of the gun and center hole of the size of the mine fuse. When this is slipped in place over the fuse, it is secured by a square piece of paper an inch or two larger than the top of the mine and with a hole punched in the middle by a wad cutter to fit over the mine fuse. Paste and press closely about the top of the mine and when it is dry it is ready for use. For stock work, they must, of course, be papered and striped, packed and labeled. Mines of saucissons are made by substituting saucissons for the stars (Fig. 44).

Devil Among the Tailors

These are made by taking a large short mine case about 4 inches in diameter and filling the bag with tailed stars, serpents and English crackers.

FIGURE 45

Devil Among the Tailors

Besides the central candle for firing it, four or more candles, one at each corner on the outside of the gun, are fastened and connected so as to burn at the same time (Fig. 45).

Anglo-Japanese Mines

These consist of a #6 mine case containing a bag filled with colored stars and Japanese or willow tree stars. Electric spreader stars also make a handsome mine. The variety of effects is almost unlimited. The genius of the artificer will suggest other combinations.

Fountains, Flower Pots and Gerbes

These are all modifications of the same principle which is a paper tube or case varying from ½ inch in diameter to 2 inches in diameter, rammed solid with one of the compositions to be given later.

Fountains

Fountains are usually from 1 to 1½ inches in diameter and 12 inches long, with a wooden point on the lower end so that they can be stuck in the ground for firing. A quarter of an

FIGURE 46
Fountain

ounce of rifle powder is sometimes placed after the last charge of composition and before the clay, both in fountains and gerbes so as to have them finished with a *bounce* or report.

Besides the regular composition with which fountains are charged, if the caliber permits, small colored stars, cut to about ¼ inch cubes and placed between the charges when ramming, greatly increase their beauty. They are then called floral or prismatic fountains. There is however some danger in ramming stars containing chlorate of potash with compositions containing free sulfur and this may be avoided by using compositions free from chlorate, such as granite stars, copper borings, etc., or perchlorate compounds.

Cascade cases are used for water falls and other designs where the fire is required to fall considerable distances to the ground. They are usually 1½ to 2 inches in diameter and 12 inches long. Where this piece is to be repeated often as at fairs, iron tubes, 2 inches in inside diameter, are sometimes used. These are stronger and can be cleaned with kerosene after using. Where the old style Niagara Falls is shown, this form of case is in general use as it saves rolling of two hundred or three hundred large cases for each display.

Flower Pots

Small choked cases ½, ⅝ and ¾ inch in diameter and from 5 to 10 inches in length, with a wooden handle in the end, provide a pretty piece of fireworks to be used by ladies and children. When properly made they are perfectly safe to fire from the hand, but this fact should be assured by first firing a few, stuck in the ground, to see that the charge is not so strong as to burst the case. The lampblack in these articles produces a peculiar effect, not entirely understood.

It may be well to mention that when ramming gerbes and the like it is advisable to begin with one charge of starting fire, especially where the composition contains steel, as they might strike fire by contact with the nipple, or, if the composition is

FIGURE 47

Flower Pot

very sharp, they sometimes burst upon starting from the depth
of the nipple allowing too much composition to become ignited
at once. Starting fire may be made from one of the less active
compositions given later.

Gerbes

These are used for all set pieces where brilliant effects or
jets of fire are desired. They are usually ¾ inch in diameter
and 9 inches long.

FIGURE 48

Gerbe

When steel filings are used in them, they should be protected,
as the saltpeter corrodes the filings and affects their brilliance.

They are rammed like rockets but on a short nipple without a central spindle.

Formulas for the three preceding items are as follows:

STARTING FIRE

Meal powder	4
Saltpeter	2
Sulfur	1
Charcoal	1

GERBES

	Formula 1[7]	Formula 2	Formula 3
Meal powder		6	4
Saltpeter	24	2	
Sulfur	4	1	
Charcoal	4	1	1
Steel filings	10	1	2

FOUNTAINS

Meal powder	5
Granulated saltpeter	3
Sulfur	1
Coarse charcoal	1
FF Rifle powder	¾

FLOWER POTS

Saltpeter	10
Sulfur	6
Lampblack	3
FFF Rifle powder	6

CASCADES
(*Niagara Falls*)

	1½ Inch Case	2 Inch Case
Granulated saltpeter	18	16
Mixed charcoal	4	4
Sulfur	3	3
Iron borings	6	7

The aluminum Niagara Falls is made from unchoked cases ½ to ¾ inches in diameter with the following mixture:

[7] Contributed by Mr. Beddy Lizza. Personal Communication. This is a very good formula and may also be used for a *volcano*.

Potassium perchlorate	2
Mixed aluminum	1
Dextrin	½

Starting fire is sometimes required; a little white star composition moistened and pushed into the end of the case is sufficient.

Volcano

This very pretty little device, easily made and perfectly safe for use by small children is produced by rolling a stout cone, 3 to 4 inches long and 2 inches in diameter at its base, on a former similar to that shown for sky rocket cones. The tip is cut off so as to allow an opening about ⅛ inch, into which a stout piece of quickmatch is inserted. A ramming mold is now made from a 4 inch block of wood, into which a tapering hole, of the same taper as the volcano cone, is worked. A case is inserted and rammed with the following composition:

Saltpeter	24
Sulfur	4
Mixed charcoal	4
Steel filings	10

A cardboard disk, somewhat smaller than the bottom of the cone, is now forced into the bottom of the volcano and secured with glue.

Wasp Light

This little contrivance is very effective and safe in destroying the nests of wasps, hornets, etc. The sketch illustrates the method of using it and the following composition is satisfactory.

Saltpeter	9
Sulfur	1½
Charcoal	5

In those cases where it is not practical to attach the light as shown, a long pole may be used. Tied to the end of a fishing pole

FIGURE 48a

Volcano

and brought into contact with a nest, it will destroy the nest without danger to the operator, as the burning composition com-

FIGURE 49

Wasp Light

pletely paralyzes the insects and they are no longer able to sting. The case should be about ¾ inch in diameter and 5 inches long.

Revolving Pieces
(*Triangles*)

These are made in various forms (Fig. 50). One form consists of a small six-sided block with concave edges on three sides into which small choked cases are fastened, either by glue, wire or nails; another consists of a triangular block on each side of which a serpent is fastened. All the serpents except the last one, must be rammed full without clay and should be primed at both ends.

A piece of paper is pasted over the joints where the two ends meet and the first one is matched for lighting. The blocks have

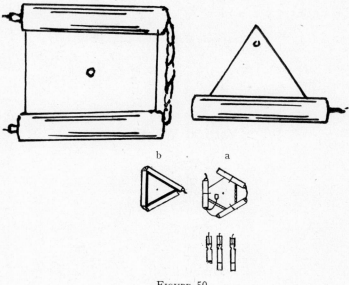

FIGURE 50

Triangles

a hole through the centers for the nail on which they revolve and
a nail is supplied with each one.

In making the larger triangles (a) take three small choked
cases ½ inch in diameter. Ram two of them with triangle composition to within ¾ inch of the end; then stop ends with ½
inch of the same composition, moistened with dextrin water, and.
ram tightly with solid rammer. The third case is closed with
clay. Now cut papering 2 inches longer than the case and cover
in the regular way. Into the choked ends of the cases after priming, twist pieces of match 1½ inches long, except for the first
one where a shorter piece will suffice. Fasten them to the block
as described above; first the one with the clayed end, then the
one with both ends open and finally the one with the short match.
Insert the match of the third case into the rear nosing of the
second one and the match of the second case into the first and

secure the joints with pasted tissue paper. On the larger triangles, a colored pot is added.

TRIANGLE COMPOSITIONS

	Formula 1	Formula 2
Saltpeter	18	12
Sulfur	2	8
Mixed coal	5	5
FFF Rifle powder	12	12

Vertical Wheels

FIGURE 51

Vertical Wheels

These are made by fastening four to eight driving cases to a wooden wheel made for this purpose. The cases are usually ½ to ¾ inch in diameter inside, either choked or rammed on a nipple with clay. They are papered and matched the same as for triangles except that connecting matches should be papered as the distance between the cases is greater than in the triangles. A little gum on the side of the case where it touches the rim of the wheel will hold it more securely than wire alone. The wooden wheels for these may be obtained from North Weare, N. H.

WHEEL CASE COMPOSITION
(*Drivers*)

	Formula 1	Formula 2
Meal powder	8	3
Saltpeter	3	2
Sulfur	1	1
Mixed charcoal	1	1
FFF Rifle powder	1	
Lampblack	½	
Steel filings, etc., ad lib.		

Saxons

FIGURE 52

Saxon

Ram two cases ½ to ¾ inch in inside diameter with a strong composition, closing both ends securely with clay, and glue them to a block as shown in Fig. 52. Holes are bored ¼ inch in diameter and just through the case as near to the clayed end as possible and at right angles to the nail hole in the center of the block on which the saxon will revolve. These holes must, of course, be on opposite sides. A piece of match is inserted into one of these holes and secured with pasted paper. Another hole is bored into the bottom of the case but on the side opposite to that of the first hole. From this one, a piece of piped match is led to the hole in the second case, fastened with a tack and well secured with pasted paper.

One or two colored pots may be attached to saxons, greatly enhancing their beauty. They are fastened to the saxon with wire so as to burn while the saxon is revolving and are matched accordingly. Also on larger vertical wheels the composition of the various drivers is varied so as to increase their effect as burning proceeds. The first case is charged with plain driving composition; the second with steel filings and the third with granite stars in mixing, etc.

SAXONS

Meal powder	4
Sulfur	2
Saltpeter	2
Mixed coal	1

Stars

This subject covers what is probably the most extensive division of the art of fireworks-making. Besides the endless variety of colors, effects, etc., we have the cut star, box star, pumped star, candle star and many others. Nearly all stars are made by dampening the composition with water (if the composition contains dextrin) or with alcohol (if it contains shellac) and pressing the caked mass into cubes, cylinders, etc., by means of various devices, described later.

Cut Stars

These are the simplest form of stars in use. Secure some dressed oak wood or maple strips, 1 inch wide and ⅜ inch thick, and from these make a frame about 12 inches wide, 18 inches long and ⅜ inch high (inside measurements). The corners should be secured by halved joints, glued and fastened with small wire nails, and clinched. Also provide a rolling pin 2 inches in diameter and 15 inches long. Now use any of the compositions given for cut or pumped stars and moisten it, a little more than for use with a pump. The most convenient way to moisten any composition is to have a large dishpan or small wooden tub into which the composition is put while water is added a little at a time. Work it in by rubbing the dampened portions between the hands until the composition is evenly moistened and a handful, squeezed firmly, will retain its shape.

Lay a piece of stiff cardboard on a marble slab, dust it with the dry composition and lay the wooden frame on it. Fill the frame with the dampened composition, pressing it down firmly with the rolling pin and leveling it off with a sliding motion so that it is flush with the top of the frame. Now, with a ruler and a table knife, cut the composition in each direction, with the cuts ⅜ inch apart, so as to cut it into cubes. This is facilitated if the frame has been previously marked at ⅜ inch intervals. Make a cut around the edge of the frame to loosen the

stars and carefully remove it. The batch is now placed in the
sun to dry. When thoroughly dried, the cubes may be broken
apart for use.

On account of the ease with which these stars ignite, as a
result of their sharp corners, they are particularly adapted to
rockets, small shells, etc., where smooth stars are apt to miss fire.
If larger stars are desired, a frame of ½ inch material or thicker
may be used.

Japanese Stars

The Japanese or lampblack star gives an effect which is always
beautiful and easily produced. It is perfectly safe under all
circumstances and has a variety of uses. The well-known willow
tree rockets and shells are made with it and it may be used as a
garniture for rockets, mines, etc. An unusual fullness is given
to any article to which a small quantity of Japanese star is
added.

Japanese stars are made in a somewhat similar manner to the
above. The great difference in weight between the bulky lampblack and the compact potash of which they are composed makes
it quite difficult to mix them thoroughly, which is necessary to
obtain good results. Furthermore, it is difficult to get lampblack
to take up water. It is necessary, therefore, to moisten it with
alcohol first. Then it will take the water more readily. The
method which the author has used with success is as follows:

	Formula 1 Ounces	Formula 2 Ounces
Lampblack	12	6
Potassium chlorate	8	4
Saltpeter	1	
Water	18	9
Alcohol	4	2
Dextrin	1	
Gum arabic		½

Mix the dextrin and saltpeter (Formula 1) together and add
sufficient water to make a gummy liquid. Heat the balance of
water to boiling and add the chlorate of potassium to it. Put

the lampblack into a large pan and pour the alcohol over it
working it in as well as possible. Now add the chlorate of potassium, dissolved in the boiling water, and stir with a stick until
it is cool enough for the hands. Finally, add the dextrin and
saltpeter. Remember that you cannot mix it too well and the
effect will be in proportion to the evenness with which this has
been done.

Take a piece of light canvas or ticking, about 18 inches square,
and put 1 or 2 handfuls of composition on it. Spread it about
1 inch thick in the center of the cloth, folding the cloth over it
and place it under a strong press. Fold up another cloth of
composition in a similar manner and place this on top of the first.
Repeat until four or five cloths are under the press and screw
it as tight as possible until the surplus water runs out freely.
Open the press, remove cakes from cloths, dry for about 2
weeks and break into pieces about ½ inch square. It is important that the lampblack be perfectly dry and free from oil to
get the best results. Sometimes it is necessary to pack a jar or
crucible with it and heat it in a bright red fire until all volatile
impurities are expelled. You will then obtain one of the most
beautiful effects in the entire art of fireworks making.

In working with Formula No. 2, sift the potash and lampblack
together several times; add alcohol; then add water in which
gum has been dissolved and proceed as for Formula No. 1.
Remember that Japanese stars take longer to dry than any other.

The best Japanese stars are made by rubbing the damp composition in a marble mortar with a wooden pestle for half an
hour.

Box Stars

Where the best and most beautiful effects are required, this
form of star is undoubtedly the most adaptable. First, these
stars burn much longer than the others; second, they are less
liable to go blind and, furthermore, they will stand more blowing from a shell than any other form of star.

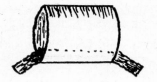

FIGURE 53

Box Star

Make some light cases of about four thicknesses of strong paper, 6 to 12 inches long on a ½ inch form. Cut with scissors into lengths of about ¾ inch. Cut some thin match into lengths of an inch or a little more. Pass a piece of match through one of the little pieces of case or *pill boxes;* bend the ends slightly around the edges as shown in Fig. 53 and dip it into a pan of composition which has been dampened as previously described. Then with the first and second fingers of the right hand press the composition as firmly as possible into it until it will hold no more. Dry in the sun for 2 or 3 days.

Pumped Stars

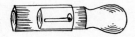

FIGURE 54

Star Pump

These are used more than any other form of star because of their regularity and the ease and speed with which they can be made. They are even more quickly made than cut stars, if the proper facilities are on hand. When only a few are required, a hand pump (Fig. 54) will serve very well. All that is necessary is to draw up the plunger, press the pump into the damp composition until it is filled and, by pressing the plunger while holding the tube, a star is ejected. When these stars are required in large quantities, however, star plates are necessary.

With these, 200 or 300 stars are made almost as quickly as one by the hand pump. The entire equipment necessary for this operation is illustrated in Fig. 55.

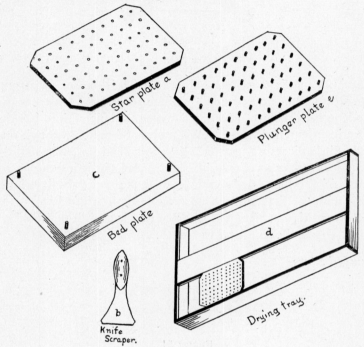

FIGURE 55

Star Plate Assembly

The standard sizes for stars are approximately as follows:

No.	Diameter, Inch	Length, Inch
1	1/4	3/8
2	5/16	7/16
3	3/8	1/2
4	7/16	9/16

The plate for making a No. 1 star must be 3/8 inch thick and have holes 1/4 inch in diameter, (a, Fig. 55). The others must be in the same proportion according to the above schedule of sizes. The plungers on the plunger plate (e) must be somewhat smaller in diameter and slightly longer than the holes in the star plate so that they will move freely and force the stars out completely. The plates are about $5\frac{1}{2}$ x $7\frac{1}{2}$ inches square. Trays (d) for holding the stars while drying should have bottoms made of brass wire netting to permit free circulation of air around the stars, enabling them to dry in a few hours. The center strip as well as the sides of the tray on top should be rabbeted so as to hold the star plate while stars are being pumped.

In order to make stars with a star plate, moisten a batch of composition with water in a dishpan, as previously described, and empty it on a rather high work table previously covered with a yard of rubber cloth. Press the plate (a) into the composition until the latter comes up through the holes. Then with the scraper knife (b) work more composition down into the holes until they appear to be full. Scrape off all surplus composition and remove to the iron plate (c), putting the side, which was previously up, down and press more composition in with the scraper. When all the holes are well filled scrape the surplus off thoroughly. Place the plate in rabbet of tray and with plunger plate (e) pump out the stars. Take care to have the right side of the plate up when pumping or the plungers will not fit the holes exactly. This side of the plate should be marked by some means.

If the plate begins to work badly on account of the composition drying on the plungers, it must be washed before being used again. The proper dampness for composition can only be determined by practice, as some mixtures require more water than others. If the composition is too dry, the stars will crumble, if too wet, they will not ignite freely. The holes in the star plate and the plungers may be much closer together than shown in the cut.

Cut, Pumped or Candle Stars

White

	Formula 1	Formula 2
Saltpeter	50	54
Sulfur	15	15
Red arsenic	15	9
Dextrin	3	3
Black antimony		15
Red lead		6
Shellac		1

Red

	Formula 1	Formula 2
Potassium chlorate	6	24
Shellac or red gum	1	3
Fine charcoal	2	4
Strontium carbonate		4
Strontium nitrate	6	
Dextrin	½	1½

Blue

Potassium chlorate	24
Paris green	9
Barium nitrate	8
Shellac	5
Dextrin	1½
Calomel, ad lib.	

Green

Potassium chlorate	6
Barium nitrate	6
Fine charcoal	2
Shellac or KD gum	1
Dextrin	½
Calomel, ad lib.	

Yellow

	Formula 1	Formula 2
Potassium chlorate	16	16
Shellac or red gum	3	3
Fine charcoal	4	¼
Barium nitrate	6	
Sodium oxalate	1	7
Dextrin	1½	1

EXHIBITION PUMPED STARS

Green

Barium chlorate	9
Shellac	1
Dextrin	¼

Red *

	Formula 1	Formula 2
Strontium nitrate	8	
Potassium chlorate	4	10
Picric acid	1½	1½
Shellac	1½	¾
Fine charcoal	1	1
Dextrin	½	¾
Strontium carbonate		3

* For hand pump; not suitable for shells.

Blue

	Formula 1	Formula 2	Formula 3 *
Potassium chlorate	48	18	16
Calomel	18	6	12
Copper, black oxide	6		
Asphaltum	6		
Dextrin	1½	1	
Paris green		4	2
Stearin		2	
Copper ammonium chloride			4
Lactose			6

* Moisten with shellac solution.

BOX STARS

Red [8]

Strontium nitrate	3
Potassium chlorate	3
Shellac	1
Dextrin	¼

Green

	Formula 1	Formula 2
Barium nitrate	3	
Potassium chlorate	4	
Shellac	1	1
Dextrin	¼	
Barium chlorate		9

[8] L. Ferrouillet. Personal Communication.

Blue

Moisten Formula No. 3 by mixing with water; Formulas No. 1 and 2, with alcohol.

	Formula 1	Formula 2	Formula 3
Paris green			25
Potassium chlorate	10		50
Potassium perchlorate		24	
Copper sulfate	3		
Copper ammonium chloride		6	
Shellac	2		
Stearin		2	8
Asphaltum		1	
Dextrin	½		5
Calomel	2		

The first of these formulas is an old one which was used before it was possible to obtain the chemicals used now. It is only suitable when the work is to be used within a few weeks. In using it, mix the copper sulfate, shellac, calomel and dextrin thoroughly; then sift and add the chlorate of potash.

In the second formula, rub the stearin and copper ammonium chloride in a mortar before adding the other ingredients.

Pink

Potassium perchlorate	16
Plaster of Paris	4
Shellac	3

Moisten with alcohol.

Yellow

Potassium chlorate	4
Sodium oxalate	2
Shellac	1
Dextrin	¼

Moisten with water.

White

Saltpeter	28
Sulfur	8
Powdered metallic antimony	6
Dextrin	1

Moisten with water.

Purple

Potassium chlorate	18
Black copper oxide	1
Calomel	6
Strontium nitrate	3
Shellac	2

Moisten with alcohol.

Lampblack Stars

These are very useful stars for shells. They are employed for producing the large aerial spiders and, since the stars take fire easily, a heavy bursting charge may be used, causing a wide spread, so desirable in this effect.

Meal powder	7
Lampblack	3
Black antimony	1
Dextrin	¼

Moisten with water, press into cakes, dry for one week and break into pieces about ¾ inch square, or use hand pump.

SILVER SHOWER

	Formula 1	Formula 2
Saltpeter	50	17
Sulfur	15	6
Red arsenic	15	
Charcoal	10	¾
Dextrin	3	
Black antimony		6
Lampblack		1

Moisten with water; press hard and cut into ½ inch cubes.

GOLDEN STREAMERS

	Formula 1	Formula 2	Formula 3
Saltpeter	8	8	25
Sodium oxalate	4		
Sulfur	2		
Charcoal	½	4	7
Dextrin	½		1
Shellac			1
Lampblack		3	
Black antimony		1	

Moisten with water and use pump.

This is a handsome star for large exhibition Roman candles.

YELLOW TWINKLER

Potassium chlorate	8
Lampblack	12
Stearin	1½
Saltpeter	1

Moisten with alcohol and shellac, and then form with hand pump.

GOLD AND SILVER RAIN

	Formula 1	Formula 2	Formula 3
Meal powder	16		4
Saltpeter	10	1	1
Sulfur	10	1	
Fine charcoal	4	1	2
Lampblack	2		
Red arsenic	1		
Shellac	1		
Dextrin	1		
Lead nitrate		3	

Moisten with water and cut into squares ¼ inch each way.

Electric Spreader Stars

The effect of these stars is quite surprising. A small pellet, no larger than a pea, will spatter over an area of 15 feet when lighted. The making of good electric spreader stars requires some care and judgment, as too much or too little dampening greatly reduces their effectiveness.

Zinc dust	36
Potassium chlorate	7½
Granulated charcoal	6
Potassium bichromate	6
Dextrin	1

Mix thoroughly all the ingredients except the charcoal and dampen until quite wet. Then add the coal, mix again and pump with the hand pump. As only coarse coal may be used, it is necessary to sift the coarse from the fine coal.

Saltpeter	14
Zinc dust	40
Fine charcoal	7
Sulfur	2½
Dextrin	1

This makes a very good substitute for electric spreader stars, for shells and rockets and is both cheap and safe to handle. It is moistened until quite wet, pressed into cakes ⅜ inch thick, cut into squares ⅜ inch each way, thoroughly dried and broken apart.

Magnesium Stars

The following is a good formula for making a star using magnesium powder:

Saltpeter	5
Magnesium powder	2

Moisten with linseed oil and pump with heavy pressure.

Steel Stars

Stars may be made with steel filings using the following formula:

Saltpeter	8
* Steel filings	2
Meal powder	1
Charcoal	1
Dextrin	¼

* Treated with paraffin.

Moisten with water and pump.

Aluminum Stars

The following formula may be used for box stars only.

Potassium perchlorate	8
Aluminum powder medium	4
Lycopodium	1

Moisten with shellac solution; fill box stars ¾ inch long and ¾ inch in diameter.

The following two most effective new stars have recently been added to our list.

FIGURE 55a

Silver Comets

SILVER COMET STARS

	Formula 1	Formula 2
Meal powder	44	8
Antimony sulfide	10	1
Mixed aluminum	3	1
Dextrin	4	1

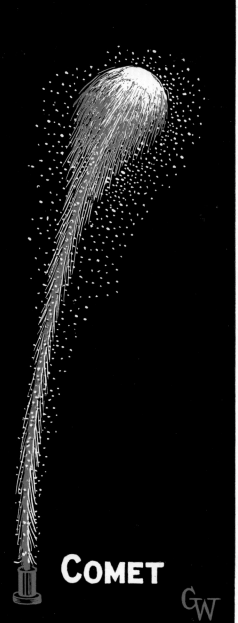

COMET

GW

GOLDEN TWINKLERS

	Formula 1	Formula 2
Meal powder	76	24
Mixed aluminum	6	3
Antimony sulfide	32	3
Sodium oxalate	8	4
Dextrin	9	2

The above two formulas give stars of exceptional beauty, but they will only function in the air, when discharged from candles, rockets or shells. On the ground they only smoulder. Also, great care must be taken to dampen them correctly.

Comets

These are large stars, about 1½ inches in diameter, fired from small paper mortars. In their simplest form, they are just large pumped stars. If the mortar is 10 inches long, a piece of piped quickmatch, 16 inches long, is bared at one end for about 1 inch and at the other about 5 inches. Lay it alongside of the comet star so that the 1 inch bared end can be bent over the bottom. Then paste a strip of paper 4 inches wide and 10 inches long and roll this around the star over the match letting the same amount project on each side. When it is dry, gather the upper extension around the match with a string and into the lower projection or bag put a half teaspoonful of grain powder and secure with a string. Now drop this in the gun and it is ready for use. Handsome effects are obtained by making half of the star of red star composition and the other of streamer composition.

A more complex comet is illustrated in Fig. 56. This is rammed into a case as shown, while the upper half, separated from the lower portion by a diaphragm of clay with a small connecting orifice, is filled with stars and a blowing charge. At the end of its flight the stars discharge giving a fine effect.

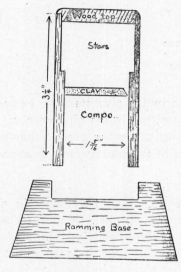

FIGURE 56

Comet

COMET STAR COMPOSITIONS

	Formula 1	Formula 2
Saltpeter	6	
Meal powder	6	3
Sulfur	1	
Fine charcoal	3	1
Metallic antimony	3	1
Lampblack		2

Aerolites

These are made by placing a comet star at the bottom of a short gun with a blowing charge but no match. Over the star is placed 2 inches of candle composition and over this 1 inch of bengal fire.

New Formulas [9]

SPIDER WEB

Saltpeter	32
Sulfur	6
Antimony sulfide	2
Coal dust	7
#36 Coal	8
A dust	4½
Dextrin	4½

Pound the composition with a 10 pound wooden stamper or stamp mill for 1½ hours. The more it is pounded, the better. This formula is for ¾ x 1 inch rammed stars.

YELLOW STAR

Potassium chlorate	17
Barium nitrate	10
Saltpeter	4½
Sodium bicarbonate	2
Red gum	5
Charcoal	1½
Dextrin	2½

Moisten until quite wet. This will work for 15 to 20 minutes. Let set and cut into cut stars.

COMETS FOR SHELLS

	Fast	*Slow*
Saltpeter	20	15
Sulfur	5	3
Mixed coal	12	8
Antimony sulfide	3	2
A dust	10	6
Dextrin	4	3

Pound for 2 hours. This formula is for a rammed star ¾ x 1 inch.

WILLOW TREE

Potassium chlorate	2½
Saltpeter	1⅜
Lampblack	4½
Sulfur	¼
Dextrin	½

[9] Contributed by F. E. Peters. Personal Communication.

Dampen with alcohol. Add gum arabic solution. Press and cut.

Wonderful Sinter Star

Potassium chlorate	32
Medium aluminum	7½
Fine aluminum	3
Barium nitrate	2
Red gum	2½
Fine coal	½
Dextrin	3

Moisten with alcohol and water. Cut stars, ½ x ½ inch and not over ⅝ x ⅝ inch.

Gold Flitter

Saltpeter	16
Sulfur	3
Fine coal	2
Sodium oxalate	4
Fine aluminum	11
Medium aluminum	4
Flake aluminum	1
Dextrin	4

This formula is for cut stars.

White Flitter

	Formula 1	Formula 2	Formula 3
Saltpeter	17	14	28
Sulfur	3		8
Fine coal	3		
Antimony sulfide	11	3	7
Fine aluminum	10	3	5
Flake	3		
Dextrin	4	1	1
Meal powder		6	3

This formula is for cut stars.

Lance Work

This is a division of pyrotechny which consists of reproducing, with colored lights, various designs, portraits, lettering, etc.
A number of wooden frames are made 5 feet wide and 10

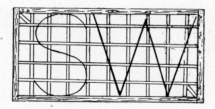

FIGURE 57

Wooden Frames for Lance Work

feet long of light lumber with outer strips ½ x 2 inches and
center strips, ½ x 1 inch, spaced 1 foot apart each way, with
braces in the two corners as shown in Fig. 57. These are laid
over the design on the floor and secured so that they do not
shift until completed. The pictures, etc., are transferred to the
frames with bamboo or rattan for the curves and light strips
of wood for the straight lines (a and b, Fig. 58).

When this has been completed, the frames should be numbered. Begin at the left-hand upper corner of the first frame and
number each consecutively to assist in getting them in their
proper places when erecting them to be burned.

Drive 1¼ inch wire nails to a depth of half an inch (b) at
intervals of 2½ inches in the curves and 4 inches on straight
lines all over the design. Be sure that there is always a nail
at every point at which two lines intersect. Now, with nippers,
cut off the heads of the nails holding the nippers at an angle
with the nail so that the place where the nail head has been cut
off will have a sharp point.

The frames are now ready for the lances. When it has been
decided what colors are to be used for the various parts of the
design, take a handful of lances of the desired color and dip
their bottoms into glue to a depth of about ⅛ inch and press
one onto each of the nails until they are attached firmly to the
cane or stick forming the design (c, Fig. 58).

When the glue has hardened, the frame is ready for matching. Take a length of quickmatch and, beginning at the upper

end of the frame, pin it from one lance to another until the entire frame is covered, following the outline of the design as much as possible (d, Fig. 58). When the end of a length of match is reached splice another to it, baring about 3 inches of the new length and slipping this bare end into the pipe of the preceding length. Secure it by tying and pasting the joint.

Leave an end of match, about 2 feet long, projecting from the lower right-hand corner of each frame so that each can be connected to the one next to it in the final assembly. Also, on one of the bottom frames leave a leader (of match) 10 to 20 feet long for the purpose of lighting the device.

Now, with a three-cornered awl make a hole $\frac{1}{2}$ inch deep through the match pipe and into the priming of each of the lances on the frame (d, Fig. 58). Then take strips of tissue paper $\frac{1}{2}$ inch wide and 3 inches long; paste a number of them onto a light board and, working along from lance to lance, secure the match to the top as shown (f, Fig. 58). Sometimes when it is desired to rush a job, to be burned the same day, the lances are secured by simply bending a pasted strip an inch wide over the top of the lance as shown at a–b, Fig. 59.

The completed frames may now be crated into lots of four, with $\frac{1}{2}$ x 4 inch strips arranged to hold them apart, for convenience in transportation.

Lances

These are the small paper tubes, from $\frac{1}{4}$ to $\frac{3}{8}$ inches in diameter, 2 to $3\frac{1}{2}$ inches long, filled with a composition burning in different colors for a duration of 1 minute. They are used for producing the different designs used in fireworks exhibitions, such as portraits, mottos, etc. The cases are rolled, and rammed, with rod and funnel as previously described.

Some lance compositions are so light that they are difficult to ram. These should be slightly dampened first. Blue lances made with Paris green and white ones using realgar are frequently used without priming as they ignite very easily.

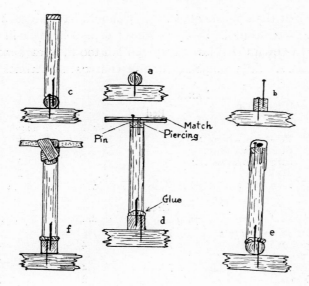

FIGURE 58

Details of Lance Work

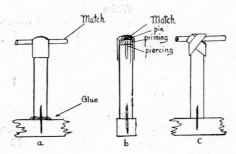

FIGURE 59

Securing Lances

A good lance should burn clearly for 1 minute, without flaring or clogging. All colors should burn about the same length of time. If a lance burns to one side it is often because the composition is not well mixed or because there is more paper on one side than on the other. They should have about 3 turns of paper all around. Badly balanced compositions will choke and burn down the sides if they contain too little carbonaceous materials. A little lampblack or fine charcoal sometimes helps.

LANCE COMPOSITIONS

Red

	Formula 1	Formula 2
Potassium chlorate	16	16
Strontium nitrate	3	
Strontium carbonate		3
Shellac	3	2
Lampblack	⅛	¼

Green

	Formula 1	Formula 2	Formula 3
Potassium chlorate	7	16	
Barium nitrate	7	4	4
Barium chlorate			5
Shellac	2	4	1
Calomel		3	2
Lampblack		⅛	
Picric acid		1	

White

	Formula 1	Formula 2	Formula 3	Formula 4
Saltpeter	9	14	5	8
Sulfur	1	4	2	2
Antimony sulfide	2			
Antimony, metallic		3	1	
Meal powder			1	
Red arsenic				1

Blue

	Formula 1	Formula 2	Formula 3	Formula 4
Potassium chlorate	20	16	12	
Potassium perchlorate				24
Paris green		5		
* Copper sulfate	6			
Copper ammonium sulfide			3	
Copper ammonium chloride				6
Shellac	4		1	
Stearin		1½	½	2
Calomel	4	3	3	
Dextrin	1			
Asphaltum				1

* See directions under *box stars* for using this.

Yellow

	Formula 1	Formula 2	Formula 3
Potassium chlorate	16	4	4
Sodium oxalate	2	2	2
Shellac	3	1	1
Charcoal	$\frac{1}{2}$		
Barium nitrate			1

In very damp climates substitute sodium metantimonate for oxalate. For amber and purple lances, the formulas given under torches may be used to advantage.

Bombshells

Bombshells represent the highest development of the pyrotechnical art and require great patience and skill for their successful production. The most wonderful pyrotechnical effects are produced by the Japanese, whereas the finest colors are made by the Europeans and Americans. Shells are made in several forms but the round ones are the most popular. Cylindrical or canister shells, however, contain several times as many stars, etc., and in some of the more complicated effects it is sometimes necessary to attach a canister to the round shell, which contains the parachute and other material (Fig. 64).

All shells are fired from mortars. The smaller ones are made of paper (up to about 3 inches in diameter) and the larger ones, of wood, copper and iron. The smallest shells with which we have to deal are the floral shells.

Floral Shells

The shells are made of hollow wooden balls which can be turned out by any wood turner. They are made in halves, usually with a rabbet to insure a close fit. Through one half drill a hole just the right size to fit snugly a piece of ordinary blasting fuse $1\frac{1}{4}$ inches long. Glue the fuse on the inside as well as on the outside of the shell case. Now fill each half with candle stars. To them add a teaspoonful of shell blowing powder and glue the edges of each half. Clap them together and

when they are dry paste a strip of paper around the place where
the two halves join. Prime the ends of the fuse (the inner end
before closing the shell). This should project through the outside of the shell about ¼ inch. Bend a piece of naked match,
about 8 inches long, around the shell in such a way that the
middle of it passes over the fuse, tacking the ends to the other
half of the shell so that they will stick over on top about 2 inches.

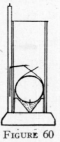

FIGURE 60
Floral Bombshell

Put an even teaspoonful of FFFF grain powder into floral shell
bottom and set the matched shell on top of it as shown in Fig. 60.
Secure with a strip of pasted paper. The larger shells require
more FFFF powder proportionately.

When it is dry, attach paper mortar with glue by slipping it
over the bottom. Measure the distance from the top of the
mortar to the top of the shell inside and mark this on the outside. Punch a hole through the mortar at this point; fit on a
top and secure. Now take a Roman candle a little longer than
the mortar and punch a hole in its side near the bottom star;
fit in a short piece of piped match. Bare the other end and slip
it into the hole in the mortar. Fasten the candle to the side with
wire and the floral shell is completed.

FLORAL SHELLS

No.	Diameter, Inches	Height of Mortar, Inches
1	2⁵⁄₁₆	9
2	2⁷⁄₁₆	11
3	3³⁄₁₆	13

Meteoric Shells

These are made somewhat differently; ¼ to ½ pound tin cans being substituted for wooden shells. They are filled with Japanese and colored stars and bursting powder in the same manner as described above. A hole is punched into the lid into which the fuse is glued. A strip of paper about 3 inches wider than the length of the can and long enough to roll around it six times is pasted all over. The filled can is placed on it and rolled up in a manner similar to that in which a case is rolled.

FIGURE 61
Meteoric Shell

The 1½ inches projecting over each end are now carefully pleated and pressed around the fuse at one end and the can bottom at the other. The shell is allowed to dry for a week before being used. The end of the fuse is trimmed and primed; a piece of quickmatch bared at both ends is laid against its side extending 1 inch beyond the fuse at the bottom of the shell. A nosing, 3 inches wide which secures the match in place, is attached to the shell and this, when dry, serves to contain the driving charge of a level teaspoonful of grain powder, after which it is gathered together and tied with twine. A dab of glue on the bottom of the bag serves to hold it on the bottom of the mortar and when this is dry it is ready for use.

Exhibition Bombshells

The principal sizes of shells used for this purpose are 4, 6 and 10 inches in diameter. For round shells, after the cases

are made as described under *cases,* the upper halves are bored
for the fuses. This may be done with a ¾ inch carpenters
brace bit boring from the inside. Fit the two halves together
accurately; bind with a strip of glued cloth and over this lay
2 or 3 layers of paper strips longitudinally, each strip overlap-
ping the one before it by about ¼ inch. If the second and third
layers are made somewhat shorter than the preceding one a
better finished job will result.

When the cases have been thoroughly dried fill them with the
desired stars through the fuse hole. When they will hold no more
add the blowing charge. The fuse should now be accurately
fitted by cutting around it with a knife ¾ inches from the top
and peeling off a layer or two of paper until it will just enter
the hole which has been made in the top of the shell for it. Glue
lower portion well and push into place until the shoulder rests
squarely against the shell case.

Attach a cloth nosing to the fuse; bare 1 inch at the end of a
length of quickmatch and attach to the bottom of the shell by
a #1 tack. Lead match up to fuse and bend at a right angle to
permit entry into the nosing. Cut piping at point of entry; insert
and secure with strong cord. The remaining match also serves
to lower the shell into the mortar (up to 6 inch sizes). Larger
shells must be provided with a separate lowering cord. The
necessary driving charge having been placed in the paper cone,
this is attached to the bottom of the shell when it is completed.

Canister or Cylindrical Shells

The making of canister shells is quite similar to that of round
ones and their construction can be readily understood from the
sketch in Fig. 62. Nevertheless, two detailed descriptions follow.

For a 4 inch diameter shell take a wooden form of 3¼ inches
diameter and 8 inches long provided with a suitable handle.
Procure some good strong paper (Kraft, sheathing or rag
board). Cut this across the grain in convenient lengths and

FIGURE 62

Canister Shell

of a width equal to the desired length of the shell. Paste it heavily with a good grade of thick flour paste and roll enough of it around the form to make a case ¼ inch thick. Do not allow the paper to soften too much before rolling. Remove from the form and dry in the shade.

Turn some pieces of maple, or similar, wood ⅜ inch thick and 3¾ inches in outside diameter and 3¼ inches in inside diameter so as to fit into the shell cases snugly. Into the piece for the top, bore a suitable hole for the fuse. Glue the piece for the bottom into the case; fill the shell as desired. Add bursting charge; glue fuse in place with surplus of glue on the inside and outside, and glue top into the shell case.

When glue is set cut some sheets of Kraft paper (24 x 36 inches—30 pound) across the grain, 4 inches longer than the shell, or a little more so that edges will lap when it is pasted into place. Soak this thoroughly with thin paste and roll about 5 layers around the shell, leaving an equal amount projecting at each end. With scissors, snip the projections into ½ inch strips longitudinally and work over top and bottom as neatly as possible so that it lies evenly and tightly around the fuse and no openings show in the bottom. When dry, it is matched and charged as described for round shells. This system is much faster than making wound shells and the author has not been able to observe any differences in their performance.

Another method, which seems to be rather extensively used by the Italians, is described thus by Prof. Tenney L. Davis in his *Chemistry of Powder and Explosives*.[10]

For a 4 inch diameter shell take a strip of bogus or news board, cut to the desired length, and roll tightly on a form without paste. When it is nearly all rolled, a strip of medium-weight Kraft paper, 4 inches wider than the other strip, is rolled in and then around the tube several times and is pasted to hold it in position. Three circular disks of pasteboard of the same diameter as the bogus tube $(3\frac{1}{2})$ are taken and a $\frac{5}{8}$ inch hole is punched in the center of the two of them. The fuse is inserted through the hole in one of them and glued heavily on the inside. When this is thoroughly dry, the disk is glued to one end of the bogus tube, the matched end of the tube being outside. The outer wrapper of Kraft paper is folded over carefully onto the disk, glued and rubbed down smoothly, and the second perforated disk is placed on top of it.

The shell case is now turned over, placed on a bench with the hole to receive the fuse and filled. The bursting charge is added. A disk of pasteboard is placed over the stars and powder, pressed down against one end of the bogus body and glued and the outer Kraft paper wrapper is folded and glued over the end.

It is thoroughly dried and wound with strong jute twine. It is first wound lengthwise; the twine is wrapped as tightly as possible and as firmly against the fuse as possible. Each time that it passes the fuse the plane of winding is advanced by about $10°$ until 36 turns have been laid on and then 36 turns are wound around the sides of the cylinder at right angles to the first winding. The shell is now ready to be *pasted in*. For this purpose 50 pound Kraft paper is cut into strips of several dimensions, the length of the strips being across the grain of

[10] T. L. Davis, *Chemistry of Powder and Explosives,* New York, John Wiley and Sons, 1943.

the paper. A strip of this paper is rubbed with paste until it is thoroughly impregnated. It is then laid on the bench and the shell rolled up in it. Stand the shell upright, fuse end up, and tear the portion of wet Kraft paper which extends above the body of the shell into strips about ¾ inch wide. Rub down smoothly, so that each overlaps the other on top of the shell and around the fuse about ½ inch. Reverse the shell and repeat the operation on the bottom. Dry outdoors when the propelling charge is attached.

A piece of piped match is laid alongside of the shell; both are rolled up without paste in 4 thicknesses of 30 pound Kraft paper, wide enough to extend 4 inches beyond the ends of the shell and held lightly in place by two strings near the end of the case. Turn bottom up; expose 3 inches of match and insert the second piece of the match in the pipe, tying with string. Introduce blowing charge of 2F gunpowder; the 2 inner layers of Kraft paper are folded down upon it, firmly pressed, and the outer layers pleated toward the center, tied and trimmed close to the string. Reverse; scrape fuse clean, in case it has been touched with paste.

Two pieces of match are crossed over the end of the fuse, bent down alongside and tied in position. The piped match which leads to the blowing charge is now laid down upon the end of the cylinder, up to the end of the fuse tube, then bent up alongside of the fuse tube, across its end, down the other side, and then bent back upon itself and tied in this position. Before it is tied a small hole is made in the match pipe where it passes the end of the Roman fuse and a piece of flat black match is inserted. The 2 inner layers of the Kraft paper are now pleated around the base of the fuse and tied close to the shell. The 2 outer layers are pleated and tied above the fuse. A 3 foot length of piped black match is now bared and an extra piece of black match is inserted and tied in place by a string about 1 inch back from the end of the pipe and covered with a section of lance tube.

Another method of matching shells is to start at the fuse by baring a half inch at the end of a match pipe and pushing this into the nosing. Bend the match ½ inch above the nosing and pass under and entirely around the shell, coming back again to the nosing. Bend once more at right angles and insert bend alongside of the starting point, first cutting through the match pipe at point of insertion. Gather the nosing closely around the match and tie as tightly as possible. This method gives a somewhat better support to the shell when it is being lowered into the mortar. Be sure to cut piping away for about half an inch where the match crosses the bottom of the shell and enters the driving charge (Fig. 63).

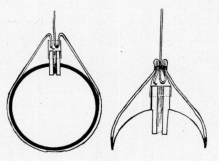

FIGURE 63
Matching Shells

Parachute Shells

These are shown in detail in Fig. 64.

Two and Three Break Shells

These are made by lightly fastening together the desired number of short canister shells with fuses not over ¼ inch long between them. The first shell has the regular length fuse. The details can be better understood from drawings than from a description. See Fig. 65.

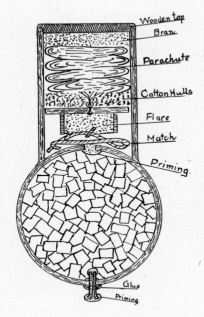

FIGURE 64

Parachute Shell

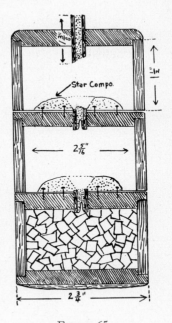

FIGURE 65

Three-break Shell

Shell Fuses

FIGURE 66

Shell Fuse

These are best made of hardware paper. Take a rod ⅜ inch in diameter (for a 6 inch shell) and a sheet of paper 6 inches wide. Paste it with thick paste all over one side and roll it up at once as tightly as possible until it has an outside diameter of ⅞ inch. The length of sheet required depends upon the thickness of the paper. When these cases are rolled, they must be dried in the shade until they are as hard as wood and rattle when struck together.

Provide a metal rammer $\frac{5}{16}$ inch in diameter, a light mallet, and some fuse composition made as follows:

Meal powder	4
Saltpeter	2
Sulfur	1

Sift and mix three times. Rest a fuse case on a firm block, scoop in a little composition and tap it with about ten light blows. Add more composition; ram again and repeat until fuse case is full. The composition in the fuse must be very hard when finished, otherwise it will blow through when used in a shell. The fuse may now be cut into the required lengths, with a fine-toothed hack saw (Fig. 66).

Length of fuses			*Inside diameter of fuses*		
4″ shell	1¼″		4″ shell	$\frac{5}{16}$″	
6″ "	1⅜″		6″ "	⅜ ″	
10″ "	1½″		10″ "	$\frac{7}{16}$″	

In some cases, a hole, ¼ inch deep, is drilled into the composition for the fuse on the end inside of the shell so that the fire from it is thrown into the shell with more force. In this case, allowance for extra length must be made when cutting the fuse.

SHELL BLOWING OR BURSTING POWDER

Grain powder	1
Meal powder	1
Saltpeter	3
Charcoal	1½
Sulfur	1

The driving charges should be coarse grain powder, the coarser, the better.

Bursting charge			*Driving charge*		
4″ shell	1¾ oz.		4″ shell	1½ oz.
6″ "	5 "		6″ "	3½ "
10″ "	16 "		10″ "	14 "

GW

ALUMINUM SHELL

An endless variety of shells may be produced. Some are listed below but the ingenuity of the pyrotechnist will suggest others as he progresses. The following are some of the well-known varieties:

Solid color shells
Variegated shells
Gold rain shells
Willow tree shells
Streamer shells
Aluminum shells
Conch shells *
Chain shells
Repeating shells
Maroon shells
Day shells

Shell Cones
(For Holding the Driving Charge)

These are made by cutting out a circle of good Kraft paper 6 to 12 inches in diameter according to the size of the shell for which it is intended. With scissors, make a cut from the edge to the center and twist it around so as to make a bag or cone of two thicknesses, pasting the edges where they meet. Put the driving charge into this and with a little paste attach it to the bottom of the shell, having previously cut the match piping where it crosses the bottom, so that fire will strike the driving charge when the shell is lighted (Fig. 67).

---

* The Conch shell consists of a 10 inch diameter round shell packed with 3 ball Roman candles made specially for this purpose. The cases of the candles are made of very strong paper so that they can be thin and no empty portion is left at top or bottom. This permits placing the greatest number of candles in the shell case. In addition to the candles, colored stars and gold rain are added, making a very effective shell.

Japanese Bombshells
(Day and Night)

The Japanese have developed this form of pyrotechny to an almost incredible degree of beauty and originality. Some of their shells are marvels of patience, skill and ingenuity.

Day shells consist of two kinds. First, those containing large figures of birds, animals, etc., made of light tissue paper sewed together like a bag and open at the bottom with a row of small weights around the edge of the bottom. The figure is folded into a small compact pile and packed into a cylindrical shell case, somewhat as parachutes are packed into rockets. When they are fired to a height of about 1,000 feet the figure is expelled with a light charge and as it falls the weights cause the bottom to unfold and the air inflates it. One of their day shells contains about a dozen paper parasols which, of course, are folded when inside the shell case but, by an ingenious construction, open as soon as the shell breaks and float to the ground much the same as parachutes do. The arrangement is shown in Figs. 66a and 66b.

FIGURE 66a
Folded Parasol

FIGURE 66b
Open Parasol

The second variety of Japanese day shells consists of colossal spiders made with smoke and vari-colored clouds. These must be seen to be appreciated. They are made by filling a round shell with smoke stars, on top of which is set a smaller canister shell containing a number of colored smoke shells, 1½ inches in diameter, and a parachute from which hangs a smoke *dragon* (Fig. 66c).

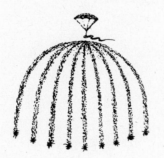

FIGURE 66c

Smoke Stars and Parachute with Smoke Dragon

The night shells include about fifty to seventy-five different effects. Up to fifty years ago, colors in night shells seem to have been unknown to the Japanese and all their devices consisted of endless varieties of tailed stars, gold and silver rain, willow trees and bright work, each one more beautiful than the other. Among some of their very unusual effects is a night shell which, upon reaching the height of its flight, throws out five red paper lanterns with a light burning inside of each one. The lanterns, when inflated, are about 2 feet in diameter and 4 feet high. When folded inside of the shell, the entire lot occupies a space of about 5 inches in diameter and 9 inches in length. Another of their original shells breaks with the customary very round effect, but only one half of the circle is filled out with stars and just the periphery of the other half is outlined. To secure this effect a wing is attached to the shell which holds it in the required position, relative to the observer, at the moment of explosion (Fig. 66d).

FIGURE 66d

A Night Shell Effect

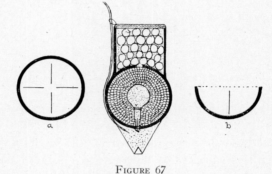

FIGURE 67

Arrangement of Smoke Stars in Shell

Smoke Stars

Smoke stars are pill boxes 5⁄8 inch in diameter and 3⁄4 inch long, closed at one end. Over the other end two strips of thin metal (copper or tin) are bent, which are secured by a paper fastening so as to restrict the opening to about one-third of its original size. The stars are charged with the composition shown under *smoke stars, yellow and olive,* the end being well primed over the metal strips. They are then arranged in the round part of the shell as shown in Fig. 67. The inside of the round shell case is scored as shown at (a and b) and by cutting half-way through the paper with a knife in order to cause the shell to burst evenly and throw the stars equally in all directions. The bursting charge, being exactly in the center, insures equal force in all directions. The little smoke shells and parachute are placed in the upper or canister portion.

Mortars

Mortars for firing pyrotechnical bombshells are made in several different ways. For shells up to 3 inches in diameter, a mortar 12 to 15 inches high, made of a number of turns of good strong paper, will serve for perhaps a hundred shots, especially if it is lined on the inside with a piece of tin or galvanized iron. If a bottom of oak or other hard wood is fitted to it and the barrel tightly wound with marline, it will be perfectly safe, cheap and light.

For 4 inch diameter shells and upward, mortars of copper tubes, shrunk one over the other so that there are four thicknesses at the bottom, three for half the length, two up to three-quarters of its barrel and one thickness for the balance, are ideal. A ring should be placed at the top.

Wrought iron tubes, wound with galvanized wire and fitted with cast iron bottoms, securely fastened by rivets or machine

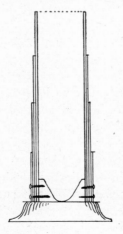

FIGURE 68

Mortar

bolts, make very serviceable guns. The bottom should be conical on the inside to accommodate the powder bag of the shell (Fig. 68).

The Japanese used wooden guns reinforced with iron bands. These are soaked in water before being used, to swell and tighten them. On account of their length, they throw a shell to a great height with a small driving charge. Half their length should be buried in the ground when they are in use. With these mortars, it was customary to pour the driving charge for shells, loosely into the mortar, drop the shell over it and fire by shaking a little dross from a port fire into the mortar. This method is very dangerous and is not recommended. Mortars with port holes on the side of the bottom like the old style military mortars are sometimes used for day shells. The cartridge of the shell is pierced with a priming rod and a piece of bare match inserted through the port hole.

Incendiary Bombs

This military device is used for setting fire to buildings, etc., in cities under attack by airplanes. After being dropped from the plane, the bombs explode upon striking any solid object in their descent.

They usually consist of a metal cylinder containing igniting materials and an explosive fired by a detonator in their points. They are provided with wings on their upper portions to maintain them in a perpendicular position in flight. They are generally about $1\frac{1}{2}$ inches in diameter and 12 inches long.

The bombs generally used in England by the Germans were made of magnesium. Upon explosion the fragments of metal take fire and burn with intense heat. A form suggested by the author consisted of three cylinders inside of one another. The outer one contained pellets of phosphorus in water; the middle one contained saltpeter and petroleum while the center held the explosive. The effect should have been something like *Greek fire*. If it was adopted by our Government is not known, but some recent communiqués reported incendiaries containing phosphorus.

Demolition Bombs

No detailed information concerning this weapon has so far been made public.

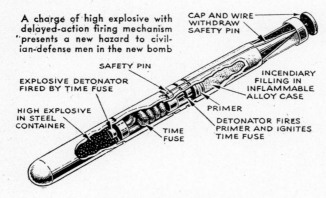

A charge of high explosive with delayed-action firing mechanism presents a new hazard to civilian-defense men in the new bomb

SAFETY PIN

EXPLOSIVE DETONATOR FIRED BY TIME FUSE

HIGH EXPLOSIVE IN STEEL CONTAINER

TIME FUSE

CAP AND WIRE WITHDRAW SAFETY PIN

INCENDIARY FILLING IN INFLAMMABLE ALLOY CASE

PRIMER

DETONATOR FIRES PRIMER AND IGNITES TIME FUSE

FIGURE 68a
Demolition Bomb

A new type of incendiary bomb carrying a delayed-action explosive charge, used extensively by both Germany and Japan during World War II.
Reprinted by courtesy of Popular Science Monthly.

Balloons

Secure some good tissue paper 20 x 30 inches (Foundrinier is best). Paste 2 sheets together on the 20 inch ends, making a sheet 20 x 60 inches. Split this lengthwise to get a sheet 10 x 60 inches. Make 12 sheets of this size, lay one on top of the other and double over longitudinally so as to have a pile 5 x 60 inches. Now, with scissors, cut along the unfolded edge as shown in a, Fig. 69, removing the shaded part. The exact line to be cut may be determined by practice until the most satisfactory shape is found. An extra sheet of heavy paper should be cut this size and reserved as a pattern, or the pattern can be made as described under *designing balloons.*

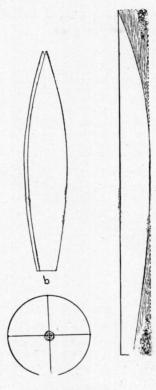

b

FIGURE 69

Balloons

Unfold the sheets; cut as above and lay one on the table before you. On top of this lay another but about half an inch
nearer to you, thus leaving an uncovered edge of the under
sheet exposed (b). Apply paste lightly to this edge and lap it
over onto the upper sheet, in this manner joining the two for
their entire length. Make six pairs of sheets like this and then
repeat the process with the double sheets until you have three
sets of 4 sheets. Join these as before, making the final closing
joint in the same manner. If the top of the balloon, where the

joints meet, is not well closed, paste a small round piece of paper over the hole.

When the balloon has dried make a ring of wire, bamboo or rattan for the bottom, with cross wires to hold the inflator (c). For a balloon of the above size, the ring should be about 15 inches in diameter. In balloons of 10 feet or more in height a wire basket is sometimes woven into the center of the ring so that an extra inflator may be added just before releasing the balloon when it is ready to rise.

Balloon Inflators

These are made in several ways. One consists of a ball of cotton wool which is saturated with alcohol or kerosene oil when the balloon is to be inflated. A more convenient inflator may be made by impregnating a ball of excelsior with paraffin and fastening it on top of cross wires of the balloon-ring. This has the advantages of being cleaner and requiring nothing further than lighting when the balloon is to be raised.

Designing Balloons

A balloon 5 feet high when inflated can be made from twelve pieces of tissue paper cut out of sheets 10 inches wide and 60 inches long. To get the proper shape for cutting these sections draw a plan of the balloon when finished, somewhat as shown in fig. 1, Fig. 70. Then make a ground plan as shown in fig. 2. Quarter the elevation plan by the two lines a and a2. The line a represents the balloon at its widest point in both plans. Line b in the ground plan is obtained by measuring the length of line b in fig. 1 from central line a2 to the edge of balloon and then taking the same distance from the center of fig. 2 and making a circle with a pair of compasses at this point. Lines c, d and e are obtained in the same manner.

Now, to make the pattern as shown in fig. 3 draw a plan of one of the sheets from which the balloon is to be cut, using the same scale as in figs. 1 and 2. Divide it by a line through its

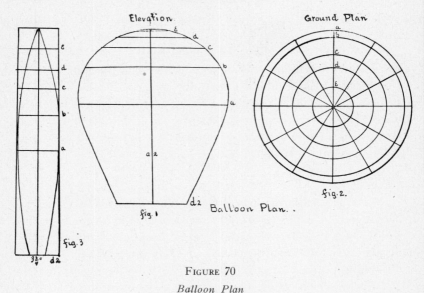

FIGURE 70

Balloon Plan

center lengthwise and then locate lines a1, b1, c1, d1, and e1 by measuring the distance from the bottom of the balloon to each cross line on fig. 1 along one edge from d2 to e. It now remains only to locate the points on fig. 3 for getting the proper shape of pattern. (a represents the balloon at its widest point in all three plans.) Now, take a pair of dividers and measure the length on line b from central perpendicular line in fig. 2 to the point where it intersects the next radial line to the right. Divide this distance equally to each side of the central line of line b in fig. 3. Do the same with lines c, d and e. On a large plan this may be more accurately done with a flexible rule, but when using dividers as above a slight allowance must be made for the curvature of the lines on fig. 2. All that is necessary now is to draw an easy line from top to bottom of fig. 3, as shown. The bottom of a 5 foot balloon should be about 15 inches in diameter. Dividing this by four will give approximately 3¾ inches for the bottom of the pattern.

Fireworks Attachments for Balloons

Fireworks make a very pretty addition to a balloon ascension and may be designed in numerous ways as the ingenuity of the pyrotechnist will suggest. A typical attachment is shown in Fig. 71.

FIGURE 71

A Fireworks Attachment

The lower portion of a gerbe is filled with red fire which burns until the balloon reaches the height of several hundred feet when the gold rain effect of the gerbe begins to function until the heading of stars, serpents, etc., is discharged. A vertical wheel, suspended from a wire and lighting when the balloon is well up in the air, makes a very interesting display.

Cannon Crackers

It would be a pleasure to omit this item entirely, as it has not only been the cause of injury to many persons but has brought ill favor upon the entire commercial fireworks industry, destroying a large business and depriving children of many perfectly harmless articles as pinwheels, sparklers, etc. However, as this work would be incomplete without a description of these articles, a discussion of them must be included.

In this field of pyrotechny the usual story of fireworks in general has been somewhat reversed. Although persons have lost limbs and life in the manufacture of Roman candles, rockets, etc., on a large scale, comparatively few serious accidents have occurred to those using them. On the other hand, although cannon crackers are one of the safest articles in the entire business to make, they have caused, during their comparatively short career in America, the loss of more hands, arms, etc., to those firing them than all other kinds of pyrotechnical devices combined.

The reason is simple. The composition of crackers is explosive only when confined or after the cracker is finished (flash cracker composition has some accidents to its debit) and the explosion of a finished cracker will not ignite others. In the case of candles, etc., a spark will fire thousands at once. When crackers are used by the inexperienced, it is difficult for them to determine whether the fuse is lighted or not. As a result the cracker explodes in the hand with disastrous results. Its bloody record has caused a number of states to legislate against its sale in sizes larger than 3 to 4 inches. However, even in these sizes, a flash cracker is powerful enough to cause the loss of a finger. Until a fuse is invented that will be consumed as it burns, this piece of fireworks will be dangerous to handle.

The first available record of the manufacture of American cannon crackers on a commercial scale was about the year 1880 when Edmund S. Hunt of Weymouth, Mass., devised a very ingenious machine into which empty cracker cases were fed from a hopper while the composition and fuse were inserted and the ends crimped in one operation. Previously, only Chinese crackers were used, but the increased loudness of the report of the American article and the reduced cost of making it soon caused it to supplant the imported one in the larger sizes. However, with the advent of the flash cracker the Chinese have again invaded the American market to a large extent.

The cases for crackers are rolled similarly to rocket cases except that paste is used only on the last turn of the farthest end of the sheet, the body of the case being rolled dry. By this means the cracker will be blown into small fragments more easily and the danger of being struck by a large piece of hard case is avoided. The fuse used is the small red cotton untaped fuse made especially for this purpose although almost any kind of blasting fuse may be used. A piece from 1½ to 3 inches long is sufficient, depending on the length of the cracker.

Various compositions are used. Those containing antimony give the loudest report, whereas those made with sulfur produce less noise. The cases should be filled about one-third with the composition to obtain the best results and the composition must be loose, not rammed. The addition of charcoal will increase the lightness of the composition and prevent its tendency to pack, but it also lessens the report.

The ends of crackers are stopped in various ways. The best way is by means of crimpers which pinch or mash the ends of the case around the fuse at one end and into a bunch or lump at the other. A daub of glue serves to retain the ends in place. Another method, which is much simpler, is to close the fuse end with clay and the other with a cork. The low grade of corks used for this purpose can be bought for less than the cost of plugs of any other sort. The corks are also held in place with glue.

To make crackers in this manner, roll the cases as directed; make a brass nipple, as shown in Fig. 72, of the diameter of the cracker desired. Drill a hole through its center, somewhat larger than the fuse so that the fuse will pass into it easily. Also provide a rammer about 6 inches longer than the cracker and drill a hole, somewhat larger than the fuse, into its lower end, and ream or countersink it a little. After setting the nipple in a block

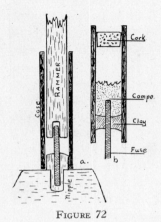

FIGURE 72

Brass Nipple Apparatus, etc., for Making Crackers

put a piece of fuse in it, slip a case on, put in enough slightly dampened clay to occupy a half inch when rammed and, with a few blows, of a mallet set it firmly.

Now remove the case and, with a sharp knife, split it open without breaking the clay and see if everything has been operating correctly, i.e., if the proper amount of clay has been used and if the fuse projects sufficiently on the inside and has not been mashed into the clay as sometimes happens if the hole in the rammer has not been made correctly on the end. The fuse should enter the hole during ramming. If this has not been done well the cracker will fail to explode.

When these matters have been properly adjusted, proceed with ramming the clay in another case and scoop in enough composition (any one from the following list) to fill the case about one-third. Then take a cork that will fit snugly, dip the small end in liquid fish glue and force it into the open end of the cracker. It is now completed and may be removed from the nipple. If too much composition is used the report will be weakened; a full case explodes very poorly.

When crackers are made on a large scale, a block of 6 dozen

nipples is used, six wide and twelve lengthwise, the same as for
Roman candles. The clay and composition may be dropped into
the lot by the use of shifting boards, the same as for candles.
Some manufacturers prefer to use a long nipple and short rammer, reversing the manner of ramming, as in this case the
cracker is rammed from the fuse end instead of from the cork
end. By this means, the danger of mashing the fuse into the
clay is avoided as the nipple on the inside protects it. However,
only the clay can be placed in them all at once by this method,
as the composition is loaded from the other end. The partly filled
cases must be removed from the spindles before this can be accomplished.

Following are the standard sizes:

No.	Length, Inches	Bore, Inches	No. in Box	Boxes in Case
Salutes				
1	2	$\frac{5}{16}$	100	20
2	3	$\frac{5}{16}$	50	25
3	3½	$\frac{3}{8}$	15	100
Crackers				
4	4	$\frac{7}{16}$	30	20
5	5	½	20	20
7	6½	$\frac{5}{8}$	10	20
9	8	¾	5	20
10	9½	$\frac{7}{8}$	3	20
12	10½	1	2	20
15	13	1¼		25*

* Not boxed.

CANNON CRACKER COMPOSITIONS

	Formula 1	Formula 2	Formula 3
Potassium chlorate	60	6	6
Washed Sulfur	23	3	2
Antimony sulfide	5		
Metallic Antimony			1
Charcoal		1	
Saltpeter	12		

If unwashed sulfur is used, the report will be considerably
louder but the danger is greater. Of the above mixings, Formula
2 is about the safest that can be made. The first gives the loudest

report. Great care must be exercised in mixing the composition for cannon crackers. Each ingredient must be sifted separately and then mixed in a tub with the fingers, preferably gloved, care being taken not to scratch the bottom with the nails.

Flash Crackers

This interesting addition to pyrotechny is one of the effects of the advent of aluminum. The following compositions may be used for flash crackers, maroons and maroon shells.

	Formula 1	Formula 2
Potassium perchlorate	50	17
Unwashed sulfur	25	5
Pyro aluminum	25	10

For Formula No. 1 mix the sulfur, and aluminum thoroughly; then add the potassium, previously sifted by itself. Mix by rolling the ingredients back and forth on a piece of paper and avoid friction of any kind.

Now prepare a block (Fig. 73) by boring several holes as shown, $\frac{7}{16}$ inch in diameter and 1 inch deep. Also prepare a nipple $\frac{3}{8}$ inch in diameter fitted into a handle (b) and some pieces of strong light paper $2\frac{1}{2}$ inches square (a). Take a piece of paper in one hand and, with the nipple in the other, press the paper around it so as to form a little cup which is now inserted in the hole in the block, pressing down until the flange of the nipple spreads the upper edges of the paper. Remove the nipple and put enough composition into the paper cup formed to fill it half full. Insert a piece of match 3 inches long; draw the paper around the match and secure tightly with two half hitches of linen twine. Remove from block, smear a little gum on one side and push into the cracker case (d) $\frac{1}{2}$ inch in diameter and 3 inches long.

When using these compositions, the ends of the crackers may be tightly closed to obtain a loud report, as directed for cannon crackers. Also, the sulfur in these mixings should not

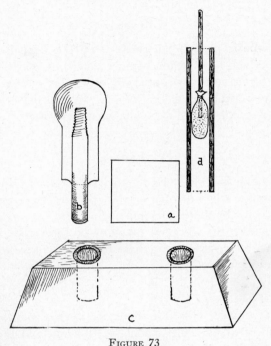

FIGURE 73

Flash Cracker Apparatus

be washed and the bags are unnecessary. Load as for cannon crackers.

A very safe and quite effective flash cracker composition, containing neither chlorate nor perchlorate of potassium, is:

Barium nitrate	4
Pyro aluminum	2
Sulfur flower	1

This will not explode by friction, concussion or spontaneous combustion. As with the other mixings, the cases must not be filled much over one-third full and all ingredients must be finely powdered and must be very dry.

Maroons

This name is probably derived from the French word for chestnuts which burst when being roasted. A maroon consists of a case of heavy paper containing an explosive charge which, when ignited, produces a loud report. Aerial maroons are arranged to explode in mid-air when fired like a shell.

Flash cracker composition may be used.

Torpedoes

By this name is understood the toy torpedoes used by children, which detonate when thrown on the ground. Most likely, these were first made by the French under the name of *pois fulminant* (mad peas) but the so-called Japanese or cap torpedoes, which constitute the largest part of those used today, are probably an American invention.

Silver Torpedoes

These are made with fulminate of silver. To prepare this, take 8 ounces of C.P. nitric acid (42%) and add 2 ounces of water gradually, stirring constantly with a glass rod. Into this put a silver dollar (or 1 ounce of metallic silver). Warm slightly until a brisk reaction takes place. When the silver is completely dissolved allow the solution to cool for 3 minutes. Then add 16 ounces of pure alcohol. Add it all at once quickly and be sure that the vessel containing the solution of the silver is quite large because a violent effervescence will take place. After it subsides add 3 more ounces of alcohol. Let stand for $\frac{1}{4}$ to $\frac{1}{2}$ hour. A white crystalline precipitate will be found on the bottom of the vessel. This is the fulminate and may be collected on a filter and dried in a shady place. A candy jar may be used for making the fulminate but a glass beaker is preferable.

The utmost care must be exercised in handling the dry powder as the slightest concussion will cause it to explode with terrific violence. A wooden spoon should be used for removing it from

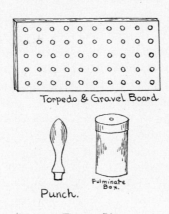

Torpedo & Gravel Board

Punch.   Fulminate Box.

FIGURE 74

Silver Torpedoes

the filter. It should be handled as little as possible and in the smallest practicable quantities. Procure a round paper box from a drug store, 1 inch in diameter and 3 inches high. Make a small hole, $\frac{1}{16}$ inch in diameter, through the cover and fill it about half full of fulminate. Now take a board 10 inches wide, 20 inches long and $\frac{7}{8}$ inch thick and, with a $\frac{3}{4}$ inch bit, bore fifty holes through it in five rows of ten holes each. This is the torpedo board. Then take a similar board and, with a $\frac{1}{2}$ inch bit, bore the same number of holes in the same position, but not quite through the board. This is the gravel board. A punch will now be required as shown in Fig. 74, the nipple being $\frac{5}{8}$ inch in diameter and $\frac{3}{4}$ inch long.

Get some tissue paper of the best grade and cut it into pieces 2 inches square. Take a bunch of these squares in the left hand and place one over each hole in the torpedo board, at the same time forcing it into the hole with the puncher so as to make a little bag. When the board is filled with paper, dip the gravel board into a box filled with gravel, tilting the board so that the surplus will run off and the holes will be just filled. Then reverse the board containing the papers and place it over the board of gravel. Hold both tightly together and turn upside down and the gravel will be emptied into the torpedo board all at once. Remove the now empty gravel board and, from the box of fulminate, shake a little of the powder into each little bag of gravel just as you would shake salt from a salt shaker. Only very little is required.

Now dip the tips of the thumb and forefinger into the paste and with the finger tips of both hands gather up the edges of the paper, bunch them together and by giving a few twists the torpedo is finished. Care must be taken not to twist too tightly or the torpedo is likely to explode in the fingers.

SILVER TORPEDOES

Name	Size Paper	Holes of Torpedo Board	No. in Box	Boxes in Case
Electric	1⅝" sq.	½"	25	50
Giant	3 " sq.	1 "	10	50

Japanese or Cap Torpedoes

These, although considerably safer to make and handle than the silver torpedoes, must be struck with much more force in order to cause them to explode. First, proceed to make the caps (Fig. 74a).

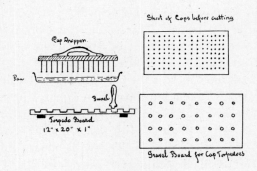

FIGURE 74a

Cap Torpedo Equipment

The equipment required is as follows: a pan 2 inches deep, 8 inches wide and 10 inches long; a number of pieces of blanket, 12 inches square, which must be well dampened before using; a cap dropper made by driving 150 8^d nails for 1 inch of their length into a wooden block 7 x 9 inches and 1½ inches thick and fitted with a handle as shown in profile in the illustration (Fig. 74a). The heads of the nails should be well levelled off so that each one touches when the dropper is placed on a flat surface. Now cut a number of pieces of poster paper 6 x 8 inches and place them in two piles on the work table. They are to receive the caps. The cap composition is made as follows:

Part 1			Part 2	
Potassium chlorate	5	oz.	Amorphous phosphorus	2 oz.
Sulfur	¼	"		
Chalk	¼	"		

Sift the ingredients of Part 1 separately; mix thoroughly in a bowl and moisten with water to the consistency of porridge. In another bowl, moisten the 2 ounces of phosphorus to the same consistency. Then, with a spoon, stir the phosphorus into the bowl containing the other ingredients. When the composition is thoroughly mixed, pour it into the pan previously mentioned. Then wash the bowls in running water.

Take the dropper by the handle and dip it into the pan of composition. Remove it and print it lightly onto the top sheet of one of the piles of poster. With a wide brush, apply a thin paste, to which a little dextrin has been added, to one side of the top sheet of the other pile and reverse it onto the sheet that has just had the caps dropped upon it. Now remove the finished sheet of caps to one of the damp blankets and repeat the operation. Place a piece of wet blanket between each sheet of caps until all the composition has been used up. Then place a light board on top of the pile of alternate caps and blankets and on this place a weight, allowing it to remain for about 1 hour. Then remove the blankets and place the pile of sheets of caps in a tight box where they cannot dry out.

Now fill the torpedo board as previously directed, using only a gravel board with somewhat smaller holes than for the silver torpedoes. Take out a few sheets of caps and with a long pair of scissors cut between each row of caps, each way, so as to separate them. Place one squarely on top of the gravel in each torpedo and, taking a handful of gravel, drop a little on top of each cap. They are now ready to be twisted as described for silver torpedoes. When several are finished they should be packed in sawdust or rice shells and removed from the workroom. Too many should not be allowed to accumulate in a pile because when they are dry the explosion of one will sometimes fire the entire lot and the flying stones may cause serious injury. Be sure never to allow the caps to become dry while making the torpedoes or in the storage box. Failure to heed this precaution has caused fatal accidents.

In making caps, when a batch has been completed, be very careful to wipe up every drop of substance that may have been spilled. Wash the pan, dropper, etc., as well as the table shears and brush well in running water and see that the washings run off.

Japanese torpedoes do not keep much over a year because the phosphorus oxidizes and, after a while, disappears entirely from the cap.

JAPANESE TORPEDOES

Name *	Size Paper	Holes of Torpedo Board	No. in Box	Boxes in Case
Am. extras	1⅝″ sq.	½″	25	100
Japanese	3 ″	1 ″	5	200
Japanese	3 ″	1 ″	10	100
Japanese	3 ″	1 ″	25	40
Cat scat	5½″	1½″	10	40
Cannon	7 ″	2 ″	10	25

* These names, of course, are arbitrary; each manufacturer uses his own brand names.

Union or Globe Torpedoes

These are made of hemispheres of clay, charged with cap composition and gravel. In general, they are made like the Japanese torpedoes but several fatal accidents have occurred in their handling and their use is discouraged.

Railway Torpedoes

FIGURE 75

Railway Torpedo

These consist of a 1 ounce ointment tin can containing a mixture similar to that used in paper caps. A strip of lead is soldered to the bottom of the box so that it can be easily attached to the rail by bending the strip around the top of it, and fires when the engine runs over the box (Fig. 75).

Other formulas are :[11]

	Formula 1	Formula 2
Potassium perchlorate	6	12
Antimony sulfide	5	9
Sulfur	1	3

Paper Caps
(For Toy Pistols)

These are made similarly to those described under Japanese torpedoes, with such variations as are necessitated by their special requirements. Brass plates with funnel-shaped holes are filled with the composition and printed onto sheets of paper. The holes should allow just enough composition to pass through at a time to form the caps. They are punched out by machinery, one sheet at a time.

[11] According to Clark, T. L. Davis, *Chemistry of Powder and Explosives,* New York, John Wiley and Sons, 1943.

Whistling Fireworks

The peculiar property of potassium picrate of whistling while burning has been known for a long time and has been made use of in the production of the amusing whistling fireworks. A more sinister use of this property of potassium picrate, that of instilling fear in non-combatants, has been found by the Germans in their *screaming bombs*.

This interesting item is made in the following manner:

In a porcelain receptacle, dissolve 1 pound of picric acid in the least possible quantity of boiling water; add ¼ pound of potassium carbonate, a little at a time, stirring continuously. When effervescence has subsided, add 1 pound of powdered saltpeter. Stir thoroughly; allow to stand for an hour and then place it on a heavy piece of filter paper in a glass funnel, to drain. When it is dry, crush to a fine powder with a wooden roller.

Although this is a reasonably safe composition, only small quantities should be handled at a time as an explosion will cause disastrous results. The dry powder may be rammed into tubes from ¼ to ¾ inch in diameter and will produce the whistling sound when burned. Bamboo tubes are most effective.

Owing to the ease with which potassium picrate detonates, whistles cannot be used in shells, but small tubes, ¼ inch in diameter and 2½ inches long, when charged with the above composition, may be placed in the heads of rockets or fastened to the outside and arranged to burn while the rocket is ascending. Attached to wheels, they are quite amusing, but the most effective use for them is in a series of six or eight, ranging in size from ¼ to ¾ inches in diameter, set side by side like a Pandean pipe and burned simultaneously.

A non-picrate whistle, safer than the one described above, is made from:

Potassium chlorate	1
Gallic acid	3

This composition makes a very good whistle and is not nearly as troublesome to prepare as the one using picric acid.

Son of a Gun
(*Spit Devil, Devil on the Walk*)

This amusing little piece of fireworks consists of a disk about 1 inch or more in diameter which, when scratched on the pavement, gives off a continuous series of little explosions, burning from $\frac{1}{2}$ to $\frac{3}{4}$ of a minute. However, they have caused some fatal accidents among small children who swallowed them because they somewhat resembled candy lozenges. Consequently their sale has been forbidden in some sections. Casualties have also resulted from the accidental dropping of cases of them. They are made as follows:

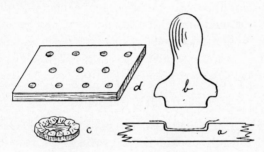

FIGURE 76

Son of a Gun Apparatus

Secure a number of boards (d, Fig. 76) of $\frac{7}{8}$ inch material and bore holes into them $\frac{1}{4}$ inch deep and $1\frac{1}{4}$ inches in diameter, as shown at a. Turn a puncher like b which will slip into the holes easily. Cut some red Foudrinier tissue paper into circular pieces $2\frac{1}{2}$ inches in diameter. Lay them over the holes on the board and punch in. Pour the composition into these and fold over the edges of the paper as in c. Permit these to set and, when hardened, they may be removed and thoroughly dried. They will then be ready for use.

Son of a Gun Composition

Mix 5 kilograms of powdered gum arabic with 5 liters of water, adding the water gradually with constant stirring. Then add 1½ kilograms of magnesium carbonate. Place this in a water bath with a thermometer arranged so that the temperature can be carefully observed and heat to 50° C. After this, add 1 kilogram of white phosphorus and stir until it is entirely melted. Continue stirring while cooling to 25° C. Then add a mixture of 3 kilograms of potassium chlorate and 2½ kilograms of red ocher and stir until a perfectly smooth product results. Then it may be poured into paper molds. Great care must be used to prevent accidents in all mixtures containing phosphorus and chlorate of potash.

Any kind of fireworks containing phosphorus (red or yellow) should be made in a separate factory, far removed from one making other fireworks. Small particles of phosphorus usually adhere to the clothes or shoes of operators. Many serious accidents have resulted from this.

Pharaoh's Serpents

The *eggs* for producing this remarkable article consist of small pellets of sulfocyanide (thiocyanate) of mercury which has the remarkable property of swelling to 25 to 50 times its original size when lighted, producing a long snake-like ash. To prepare it, make a concentrated solution of mercuric chloride and add, little by little, a solution of potassium sulfocyanide, stirring constantly. A grayish precipitate will be formed and when the last addition of sulfocyanide no longer produces cloudiness, permit the mixture to settle. Drain the supernatant liquid off as much as possible, remove the precipitate to a filter paper, placed in a glass funnel, and wash slightly. When it is thoroughly dried, reduce it to a fine powder. When ready to form the *eggs,* moisten the composition very sparingly with a weak solution of gum arabic to which may be added a pinch of saltpeter and, with the appliance shown in Fig. 77, form cones.

Plate for mass production.
For single pills.

FIGURE 77

Pharaoh's Serpent Egg Form

Magic Serpent
(*Black*)

This device, invented by the Germans, produces an immense, long, black snake. It is quite similar to the Pharaoh's serpent but is in no way chemically related. It is made of the following:

Naphtha pitch	10
Linseed oil	2
Fuming nitric acid	7
Picric acid	3½

Reduce the pitch to a fine powder; add linseed oil and mix well in a mortar. Add the fuming nitric acid, always a little at a time. Allow to cool for 1 hour. Wash several times with water, the last time allowing the mass to stand in the water for several hours. Dry thoroughly; powder finely and add picric acid, rubbing it in well. Moisten with gum arabic water and form into pellets about the size of a #4 star.

It appears that naphtha pitch can be obtained only in Germany, and there only with considerable difficulty. A fairly good article may be made by melting together equal parts of

Syrian asphaltum and roofing pitch. Add 5% stearin to the final product and form pellets.

This article has also been used in innumerable ways, such as hats, barrels, boxes, groups of a number of snakes which burn simultaneously, etc. It is a very entertaining piece of fireworks and is said to be non-poisonous whereas the Pharaoh's serpents

FIGURE 77a

Magic Serpent

are toxic and the fumes given off must not be inhaled while they are burning.

Prof. Tenney L. Davis of the Massachusetts Institute of Technology has made an exhaustive study of this interesting contrivance and with his kind permission the following extract of his work is given.[12]

"Thirty grams of naphthol is mixed intimately with 6 grams of linseed oil, and the material is chilled in a 200 cc Pyrex beaker surrounded with cracked ice. Twenty-one cc of fuming nitric acid * is added in small portions, one drop at a time at first, and the material is stirred, kneaded, and thoroughly

[12] T. L. Davis, *Chemistry of Powder and Explosives,* **Vol. 1,** New York, John Wiley and Sons, 1943.
  * Density 1.50.

mixed at all times by means of a porcelain spatula. The addition of each drop of acid, especially at the beginning of the process causes an abundance of red fumes, considerable heat and some spattering. It is recommended that goggles and rubber gloves be worn, and that the operation be carried out under a hood. The heat of the reaction causes the material to assume a plastic condition, and the rate of the addition should be so regulated that this condition is maintained. After all the acid has been added the dark brown, doughlike mass becomes friable on cooling. It is broken up under water with the spatula, washed thoroughly and allowed to dry in the air. The product is ground up in a porcelain mortar with 10.5 grams of picric acid, made into a moist meal with gum arabic water, pelleted and dried. A pellet ½ inch long and ⅜ inch in diameter gives a snake about 4 feet long which is smooth skinned and glossy. It has luster like that of coke, is elastic and spongy within."

If it is impossible to secure naphthol-pitch, which is a by-product in the manufacture of β-naphthol, a fairly good sub-

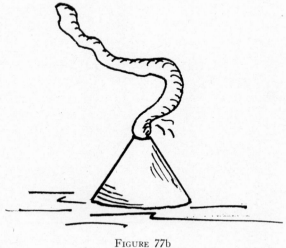

FIGURE 77b

Snake Nest

stitute is made by melting 4 parts of roofing pitch together with 6 parts of Syrian asphalt which has been reduced to a fine powder when cold.

Snake Nests

(*Snake in the Grass*)

These consist of small cones of tinfoil containing a preparation which, when ignited, produces a grass green pile of ash from which emerges a Pharaoh's serpent.

Cut some tinfoil into circles 1½ inches in diameter. Cut these again from the periphery to the center as shown in a, Fig. 78. Fold them around the former (b) so as to make little cones (d) and insert into a block (c), filling them with the following composition:

Saltpeter	1
Ammonium bichromate	2
Dextrin	1

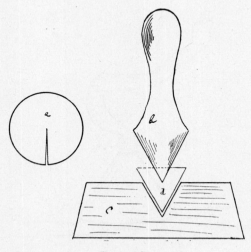

FIGURE 78

Snake Nest Tools

When the cone is quite full (up to the edge of the block), press a pellet of Pharaoh's serpent powder into the center. Fold over the edges to the center and remove from the block. When using, light at point of cone.

Colored Flames

These are made by dissolving various substances in alcohol. A copper can filled with cotton is impregnated with the alcoholic solution. It is lighted by a tuft of cotton protruding from the opening (Fig. 79).

FIGURE 79
Colored Flame

For:	Use:
Green Flame	Boric acid
Red "	Strontium or lithium chloride
Yellow "	Sodium chloride
Blue "	Copper sulfate or cesium carbonate

Before lighting, sprinkle a little of the powdered chemical over the cotton where it projects from the can.

Colored Fire Sticks

These consist of thin wooden sticks similar to the applicators used by physicians for applying iodine, etc. They are dipped for half their length into colored fire compositions, in a more or less liquid state.

One method is to melt one part of shellac in an iron pot and

FIGURE 80

Colored Fire Sticks

stir in five parts of finely powdered strontium nitrate. To keep this sufficiently liquid it must be kept quite hot by the use of a steam kettle. This is for red sticks. Another method is to dissolve the shellac in alcohol and add the strontium. The consistency of the mixture can be easily regulated by using more or less alcohol as required. After dipping, the sticks are dried and are ready for use.

Green color is not easily achieved, barium nitrate being substituted for the strontium, with the addition of a little calomel. Nine parts of barium chlorate and one part of shellac dissolved in alcohol may be used. A small amount of lampblack improves the burning but detracts from the color, especially the green. The sticks are pushed into a groove in the bar as shown in Fig. 80 for dipping and drying.

Ruby and Emerald Shower Sticks

These are much more effective and are made in a manner similar to the above, using the following compositions:

RED		GREEN	
Strontium nitrate	6	Coarse aluminum	6
Fine aluminum	6	Barium chlorate	4
Potassium perchlorate	2	Shellac	3/4
Shellac	1		

Dissolve the shellac in alcohol and add the other ingredients, previously well mixed. Stir thoroughly to a consistency of thick glue and dip sticks which were previously arranged in holder so they may be placed on a rack to dry.

Sparklers

These are made in the same general way as the above. In effect, they throw off a shower of beautiful sparks. There are several varieties of sparkling sticks which are sold under this name. The principal one consists of pieces of wire or thin twisted metal, part of which is covered with a composition containing steel filings.

The Japanese make a similar article with the composition in a twisted paper, but to make this requires a degree of skill which western races do not seem to have attained.

STEEL SPARKLER COMPOSITION

	Formula 1	Formula 2
Sulfur		4
Fine steel filings	12	8
Fine aluminum powder	1	1
Fine charcoal		6
Potassium perchlorate	6	
Dextrin or gum arabic	2	
Saltpeter		20
Shellac solution		Enough

STEEL SPARKLER [13]

Saltpeter	64
Barium nitrate	30
Sulfur	16
Coal	16
Antimony sulfide	16
Powdered aluminum	9
Dextrin	16

The steel must be protected from corrosion with paraffin. The gum should be made up of the consistency of mucilage. Mix the ingredients thoroughly and add gum solution until a mixture is obtained that will adhere to the wires when they are dipped into it. The dipping time varies in different sections and with different runs of ingredients. In practice, bunches of wires

[13] According to Clark. T. L. Davis, *Chemistry of Powder and Explosives,* New York, John Wiley and Sons, 1943.

are dipped at once and slowly withdrawn in a current of warm dry air which causes the mixture to adhere evenly.

A sparkler of great brilliance, which is very effective is easily made as follows:

Take 3 pounds of dextrin and add 20 pints of water, a little at a time, stirring continuously in order to avoid lumps. Mix intimately 10 pounds of potassium perchlorate with 7 pounds of finely powdered aluminum and add this to the gum water, stirring until a perfectly smooth mixture is obtained. Wooden sticks may now be dipped into it to the desired depth while it is contained in a deep vessel and placed in a suitable rack for drying. It may be necessary to dip the sticks several times, depending on how much composition adheres to them. They should be dried with the composition end up, the first time, so that not too much composition accumulates on the point.

Water Fireworks

Water fireworks (Fig. 80a) consist mainly of five or six varieties as follows:

1. Floating gerbe or Roman candle. A cone-shaped piece of light wood is bored with a hole of suitable size to take the gerbe or candle as shown. In order to secure an upright position with Roman candles, it is sometimes necessary to place a charge of iron filings or lead shot in the bottom of the case.

2. Floating tableau lights. These are merely colored pots placed on a suitable board.

3. Diving devils. A sharp gerbe is fitted with a hollow head set at an angle with the case. Careful adjustment must be made in order to insure proper floating of the gerbe which will cause it to dive into and come up out of the water properly. This is perhaps the most amusing piece of water fireworks as well as the one calling for the most careful work. The tip of the float must be weighted so as to cause it to dive and yet be buoyant enough to make it rise again.

4. Fish. They are made similar to the diving devil except

that not so much adjustment is necessarry since they only *run* around on top of the water.

5. Water wheels. These are ordinary vertical wheels set on a board float as shown.

6. Floating mine. The construction of this is self-evident. The fish and devils should be heavily coated with paraffin when finished; even the nosing of the match should be protected in this manner and a waterproof fuse, properly primed, should be used for lighting.

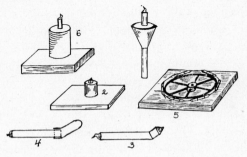

FIGURE 80a

Water Fireworks

Water fireworks are only practical on quiet ponds and small lakes and are usually fired from a skiff. Great care must be taken to protect the supply in the boat from sparks given off by those being burned, in order to prevent accidents to the operators.

Smokes

Smoke Screens

Although these are really no part of the pyrotechnical art they do come under the heading of *military pyrotechnics*. They often consist of a fine stream of titanium tetrachloride which is sprayed from an airplane at a suitable height and which, in falling, produces a dense smoke intended to screen what is behind it. The liquid is projected from the plane at the same rate as the plane's forward movement through the air, so as to cause the droplets to fall perpendicularly.

The army is said to use another type of smoke screen, the action of which is a military secret.

Colored Smokes

This branch of pyrotechny seems to have been somewhat overlooked, though its possibilities for daylight entertainment as a supplement of night displays could open an interesting field for those with enough imagination to develop it.

There are as many colors and tints of smoke as there are flames and aerial combinations. The simplest form of smoke used in pyrotechny is the *smoke pot,* as used in spectacles like *The Last Days of Pompeii* and *Burning of Rome* where it is desired to give the effect of destruction by fire. Smoke and spark pots consist of short mine cases about 4 to 6 inches in diameter and 6 to 12 inches in length. A basic formula subject to variation is:

SMOKE POT

Saltpeter	4
Lampblack	1
Charcoal	1
Realgar	1
Rosin	1

This composition is rammed into a large, partly choked case, about half the length of a gerbe. A starting fire as given below is also necessary.

Saltpeter	6
Sulfur	1¼
Antimony sulfide	1
Meal powder	1

About ¼ inch of this is placed on top of the other composition before putting the top on the case. This may consist of a wooden disk with a hole, fastened in place with small nails. A tin can may be used as a container.

SPARK POT

Meal powder	2
Fine charcoal	1
Sawdust	1

The following are good formulas for smoke compositions:

WHITE

Sulfur	16
Saltpeter	12
Fine charcoal	1

Use a little white star composition for *starting*.

	Formula 1	Formula 2
Zinc oxide		22
Zinc dust	2	28
Hexachloroethane	1	50

Use starting fire.

BLACK

	Formula 1	Formula 2	Formula 3
Magnesium powder		18	1
Hexachloroethane	24	60	3
Alpha naphthol	6		
Anthracene	2		
Aluminum powder	4		
Candle composition	6		
Naphthalene		21	1

Use white star composition for starting.

STARTING FIRE

Saltpeter	14
Sulfur	4
Red arsenic	4
Dextrin	1

Pyrotechnics

BRIGHT RED

	Formula 1	Formula 2
Potassium chlorate	1	
Potassium perchlorate		5
Lactose	1	
Paranitraniline red	3	
Antimony sulfide		4
Rhodamine red		10
Gum arabic		1 [14]

DARK RED

Potassium chlorate	7
Lactose	5
Auramine	2
Chrysoidine	6

CANARY YELLOW

Potassium chlorate	1
Lactose	1
Paranitraniline yellow	2

Use red star composition for starting.

OLIVE YELLOW

Saltpeter	1
Red arsenic	1
Sulfur	1
Antimony sulfide	1
Meal powder	1

No starting fire is necessary. This is suitable for smoke stars.

GREEN

	Formula 1	Formula 2
Potassium chlorate	6	
Potassium perchlorate		6
Lactose	5	
Auramine	3	
Indigo, synthetic	5	
Antimony sulfide		5
Malachite green		10
Gum arabic		1 [15]

Use star composition for starting.

[14] According to Clark. T. L. Davis, *Chemistry of Powder and Explosives,* New York, John Wiley and Sons, 1943.

BLUE

	Formula 1	Formula 2
Potassium chlorate	7	
Potassium perchlorate		5
Lactose	5	
Indigo, synthetic	8	
Antimony sulfide		4
Methylene blue		10
Gum arabic		1 [15]

Use colored star composition for starting.

Almost any pastel shade desired may be produced by combining the above formulas containing aniline dyes.

Recently, during the last war, great advances have been made in the production of colored smoke. A hundred years ago, the Japanese made a thick yellow smoke by the use of arsenic and antimony. In World War I, some more or less successful smokes were made by the use of aniline dyes. However, it was not until World War II that this interesting process was brought to a high degree of development.

Smokes of every color and shade are now available for both military and spectacular work. Black and white smoke are still made from a hexachloroethane base, of very high efficiency, using zinc dust for white and naphthalene with powdered magnesium for black. As these are somewhat difficult to ignite, starting mixtures are necessary.

At present, the colored smokes are all made by the volatilization of aniline dyes, using a heating mixture of potassium and sugar, with sodium bicarbonate as a restrainer to control the combustion and prevent flaming, in which case the smoke is destroyed. The following compositions operate so easily that no special container is necessary and they will even function if burned as a loose powder and no starting fire is required. The anilines should be very finely powdered. Most of their bulk should pass through a 200 mesh sieve.

[15] According to Clark. T. L. Davis, *Chemistry of Powder and Explosives,* New York, John Wiley and Sons, 1943.

The Chemical Warfare Service has developed the following formulas:

Yellow #1

Auramine 0	38.0
Sodium bicarbonate	28.5
Potassium chlorate	21.4
Sulfur	9.4

Yellow #2

β-Naphthalene azodimethylaniline	50.0
Potassium chlorate	30.0
Sugar	20.0

Red #1

Diethylaminorosindone	48
Potassium chlorate	26
Sugar	26

Red #2

Methylaminoanthraquinone	42.5
Potassium chlorate	27.4
Sodium bicarbonate	19.5
Sulfur	10.6

Orange #1

1-Amino-8-chloroanthraquinone	39.0
Auramine O	6.0
Sodium bicarbonate	24.0
Potassium chlorate	22.3
Sulfur	8.7

Orange #2

α-Aminoanthraquinone	24.6
Auramine O	16.4
Sodium bicarbonate	23.0
Potassium chlorate	25.9
Sulfur	10.1

Violet

1-Methylaminoanthraquinone	18.0
1,4-Diamino-2,3-dihydroanthraquinone	26.0
Sodium bicarbonate	14.0
Potassium chlorate	30.2
Sulfur	11.8

GREEN

Auramine O	11.7
1,4-Di-p-toluidinoanthraquinone	28.3
Sodium bicarbonate	24.0
Potassium chlorate	25.9
Sulfur	10.1

BLUE

1,4-Dimethylaminoanthraquinone	50.0
Potassium chlorate	25.0
Sugar	25.0

Smoke Grenade

The use of colored smokes in warfare has been developed to a high degree of efficiency. For daylight signalling over considerable distances smoke clouds of various colors provide a means of communicating in cases where no other system may be used. For this purpose a *smoke grenade* has been produced as illustrated in Fig. 81.

The container consists of a metal can 4¾ inches high and 2¼ inches in diameter. It is charged from a ½ inch hole in the bottom. The top contains four holes, ⅜ inch in diameter (1 and 2 shown in sketch) through which the smoke, produced by the burning composition, is permitted to escape. The combustion of the ingredients is incomplete and, for this reason, the dense smoke is produced. If they burned more completely only almost invisible gases would result. The four holes in the top of the can restrict the combustion and only sufficient heat is generated to vaporize the anilines instead of burning them.

The ignition of the contents of the can is brought about by a very ingenious device similar to the one on regular hand grenades and which renders them harmless until ready for action. The large tube entering the can contains some of the composition and into this are forced four holes about ⅛ inch in diameter filled with a much more easily burning composition, e.g., starting fire. This projects up to and around the detonator and catches fire when the hammer explodes the fulminate in the

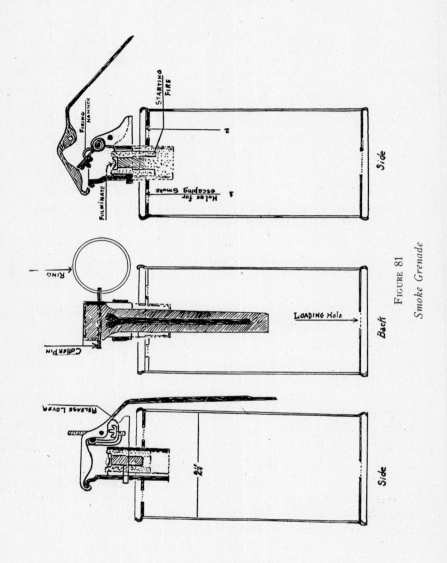

FIGURE 81

Smoke Grenade

198

firing mechanism, as shown in the figure. It, in turn, communicates the fire to the smoke composition which is more difficult to ignite. However, sometimes starting fire may be omitted as some of the mixings take fire quite easily.

When using, grasp the can firmly in the hand, being sure to keep the release lever tightly pressed against the side of the can. Remove cotter pin by pulling ring and throw grenade a distance of 15 feet in the direction in which the wind is blowing.

As soon as the lever is released, which occurs when the can is thrown, a spring causes the lever to fly off, freeing the hammer which is actuated by the same spring. The hammer strikes the fulminate which fires the starting composition and eventually the smoke mixture.

The formulas given at the beginning of this article are suitable.

Smoke Shells
(*Smoke Clouds*)

These are usually made by filling small shell cases (not over 3 inches in diameter) with a finely divided powder of the desired color of cloud to be obtained. To the inside end of the shell fuse is attached a small bag of gunpowder which should be located as near the center of the shell as possible. This, when exploding, serves to scatter the colored matter and produce the cloud. The arrangement of the fuse as shown in Fig. 67 may be used.

For red, use American vermilion powder; for blue, ultramarine powder; for green, Paris green; for yellow, chrome yellow; for white, chalk; and for black, ivory black.

Smoke Cases

For producing a dense smoke of the desired color, a paper tube, 1 inch in inside diameter and 4 inches long, is desirable. Into this are bored four or five holes, ¼ inch in diameter, on a spiral line. Both ends of the case may be closed with clay or wooden plugs. Do not pack smoke compositions; ram very lightly.

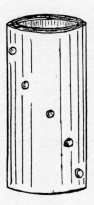

FIGURE 81a

Smoke Case

Part IV

Exhibition Fireworks

The following sketches of exhibition pieces are largely suggestive. While they cover almost the entire gamut of pyrotechnical devices except lancework, shells, etc., they are subject to endless variations, and yet the basic principles are practically all shown. They may be made larger or smaller, simpler or more elaborate, according to the importance of the intended display.

Snake and Butterfly

Originally described as the *Salamandre,* this ingenious device appears to have been originated by Ruggiere in 1737 at Versailles. It has since been frequently displayed by the C. T. Brock Co. of Surrey, England, at the late Crystal Palace.

It consists of a snake squirming around in the air after a butterfly which manages to evade it. The framework consists of an endless chain of wooden links 4 x 8 inches, bolted together and running on four sprockets and four idlers of a suitable size as shown in Fig. 82. When mounted, a crank is attached to the rear of one of the sprockets, by which the whole is operated. The snake and butterfly are made of lancework which is attached to the chain.

Rocket Wheel

This is a very old, yet always attractive, device. It consists of two wheels, 3 feet in diameter, attached to opposite ends of an axle arranged to revolve horizontally on a spindle as shown. The rockets pass through screw eyes along the rim of the wheel and are matched to fire at intervals as the wheel revolves, by

201

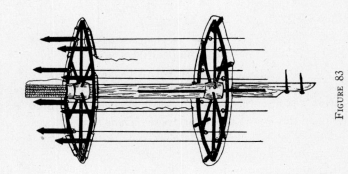

FIGURE 83
Rocket Wheel

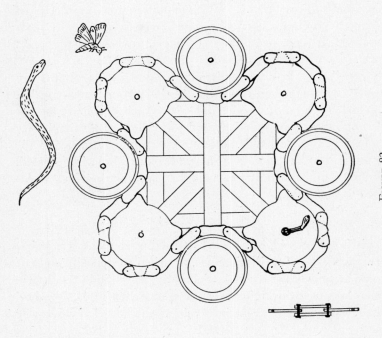

FIGURE 82
Snake and Butterfly

being connected to successive drivers. On the top is a battery of Roman candles. The top wheel is fitted with ordinary drivers containing steel filings and matched to burn two at a time, one each on opposite sides. The lower wheel is fitted with aluminum gerbes which burn simultaneously with the last two drivers of the top wheel. The latter are set at an angle with the axis of rotation so as to give a wider spread of fire. The battery of candles starts with the second pair of drivers of top wheel (Fig. 83).

Revolving Globe

This simple, yet baffling, device is constructed as shown in Fig. 84. The frame may be secured in different sizes, all ready for the lances, etc., from manufacturers of fireworks wheels, in North Weare, N. H., or it may be constructed by the pyrotechnist himself according to suggestions given in the sketch.

When the piece is burning, the globe appears to be revolving, first in one direction, then in the other, in a most amusing manner.

Living Fireworks

These were invented by the Brock Co. and consist of men wearing asbestos suits and having light frames attached to their bodies, which are covered with lancework. While these are burning the body movements are portrayed in fire.

Changing Pictures

This intriguing pyrotechnical device starts with one picture and, while burning, dissolves into another. For instance, a piece of lancework starts as a floral design. When it is half burned out, it changes to a portrait, etc.

The portrait is sketched first and marked out for full-length

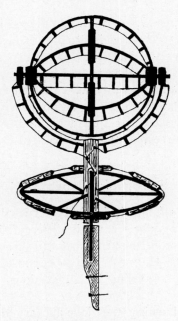

FIGURE 84
Revolving Globe

white lances. Other lines are now drawn into the design which make of it a floral picture. These supplemental lines are covered with half-length lances in colors. When the piece first burns the complete picture shows but as the short lances burn out the portrait suddenly appears.

Mosaics
(*Feu-croisés*)

These are set pieces in which gerbes are placed at right angles so as to produce the effect of squares of fire, usually with saxon wheels between to enhance the effect.

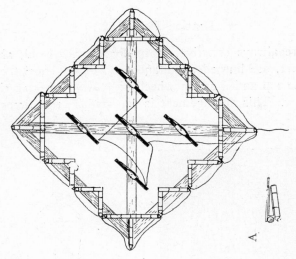

A

FIGURE 85
Mosaic
Where two gerbes meet while burning one should
be raised at its front end as shown at A.

FIGURE 85a
Mosaic when Burning

Lattice Poles

These are always effective, especially when several, about 10 feet apart, are burned simultaneously. They consist of a gerbe at the top and a saxon wheel with two gerbes at right angles below. This arrangement is repeated several times on each pole.

FIGURE 86

Lattice Poles

Saxon Cross

This device consists of four saxon wheels arranged on a square frame, with two gerbes set at 45° angles below and above each one, so as to form the diagonal square when burning.

FIGURE 87

Saxon Cross

Revolving Sun

This piece is built on a wooden frame made in the shape of a wheel, usually of six or eight spokes, the outer rim of which consists of two rows of stout cane about 4 inches apart, onto which drivers are attached at a sufficient angle to cause the wheel to revolve when burning. In the center may be placed any number of saxons according to the diameter of the wheel.

Caprice

This is always an effective device, forming a double fountain. This may also be made in various sizes; the larger, the more beautiful.

FIGURE 88

Revolving Sun

Beehive

This piece never fails to get applause. A round beehive, about 10 feet or more in height is made of lance work. Three serpent mines, pointing slightly forward, are attached to the center of the frame and matched to lances so that the first one burns when the piece is well lighted; the second when half burned and the last one just before the piece is burned out.

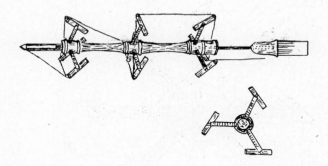

FIGURE 89
Caprice

Chromatropes
(*Guilloché*)

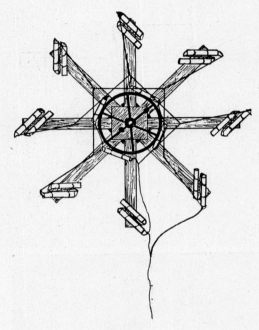

FIGURE 90

Chromatropes

These are made by placing drivers at an angle of 45°, either single or double, on the ends of two wooden crosses—and each four pairs are so arranged as to revolve on a spindle in opposite directions. There is a vertical wheel in the center.

Yule Log

Various means have been employed to produce colored flames when yule logs are burned in open fireplaces. The best results were obtained by the author in the following manner:

Secure a well dried log of any porous wood (pine and oak are

FIGURE 90a

Chromatropes Burning

not so good) and paint it as heavily as possible, all over, with a mixture of finely powdered copper-ammonium chloride in shellac solution. When this is thoroughly dry and the log is rolled into a bed of embers, it will give off beautiful green and deep-blue flames for quite a while.

Another method consists in impregnating sawdust with the salt. When dried and thrown into a wood fire, the same effect is produced. For red flames use strontium nitrate and for green flames barium nitrate.

Niagara Falls

The directions given here were in use before the advent of aluminum in pyrotechny. To many of the older workers, this

Falls, with its quiet dignity and deep characteristic roar, was more effective than the present somewhat garish display with aluminum. The method for producing either one has been given.

Part V

MISCELLANEOUS

Lycopodium Pipe

When it is desired to produce a sudden flash of flame, use is made of lycopodium, an inflammable powder consisting of the spores of certain moss-like plants. When a stream of this powder is projected across an alcohol flame, it takes fire in the air with startling results.

Years ago, before modern lighting effects were introduced into the theater, this was the only means of simulating lightning. In the same way, when the devil appeared on the stage, he was always preceded by a big flame. Although, much more realistic than the present devices, its use was discontinued, it was claimed, on account of the fire hazard. However, this idea was fallacious for the reason that there is very little heat generated by the burning of lycopodium.

The best results are obtained by the use of the lycopodium pipe shown here. The bowl of the pipe is filled with the powder through the opening in the center into which fits the receptacle containing some cotton wool which is impregnated with alcohol. When air is blown into the stem of the pipe an immense flame issues from the bowl. Where the pipe is to be used very much a bellows is attached to the pipe stem.

Water Light

This is a marine appliance used for assisting a person who has fallen overboard at night to see the life buoy which is thrown to him to support him until the life boat can pick him up.

213

FIGURE 91

Lycopodium Pipe

It consists of a 1 pound tin can containing pieces of reddish-brown calcium phosphide which, upon being thrown into water, evolves hydrogen phosphide gas. This gas takes fire spontaneously when coming in contact with air. The can has a ring on one end by which it is fastened with a small rope to the life buoy. A piece of perforated metal at the other end prevents the contents from falling out when the cover is removed.

As soon as someone falls overboard the cover of the can is

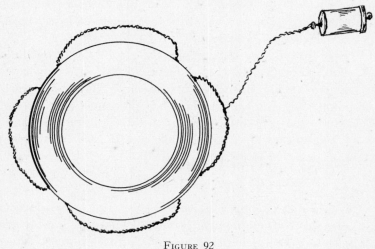

FIGURE 92

Water Light

torn off and the buoy, with the can attached, is thrown over-board as near as possible to the one who is in the water. As the calcium phosphide comes in contact with the sea water, bubbles of gas are formed and as these reach the air they take fire with a bright yellow flash. This guides the one in the water (if he can swim) as well as the boat to the spot where the buoy is.

Chinese Firecrackers

As far as it is known to the writer, a detailed account of the manufacture of this interesting little article of pyrotechnics has never been given in English. There are undoubtedly more of these made than of any other piece of fireworks. The ingenuity of the Chinese in the production of unbelievably large quantities of these firecrackers is only equalled by the many other unusual things done by this most patient and painstaking race.

Figure 93

Chinese Woman Making Firecrackers

The annual imports of Chinese crackers to this country alone amounts to three million dollars, which, divided among the various sizes, would amount conservatively to eight billion crackers.

The tubes or firecracker cases are 1¾ inches long, ¼ inch in outside diameter and have a bore of ⁵⁄₃₂ inch. They are rolled of a grade of paper unknown in this country, perhaps the lowest grade of paper made, unsized and quite irregular in character, a sort of coarse blotting paper. A small amount of gum water or rice paste is used as a binder and the case is finished with 1 turn of very thin red, green or yellow paper. They are rolled in lengths of 1 to 2 feet and then cut to the required size.

Now a block is prepared for gathering about 1,000 of these tubes into a hexagonal bundle, as follows; a piece of hard wood about 1 inch thick and cut into a hexagon, with sides 5 inches wide, is provided with pointed wooden or metal pins, each ¾ inch long and ⁵⁄₃₂ inch in diameter, set into the wood base so that the top ¼ inch projects and they are exactly ¼ inch apart.

FIGURE 94

Chinese Boy Making Firecrackers

They are also arranged in a hexagon with 4 inch sides. A tube is then slipped over each pin until the entire block is filled, a wooden frame the same size as the outside of the block, ½ inch thick and having an inside diameter slightly greater than the assembled tubes so as to be able to slip snugly around them having previously been provided. This is slipped up and down a few times to shape the bundle nicely and a string tied around it to secure it further.

A piece of white paper is now pasted over the top of the bundle. When dry it is removed from the form and a piece of paper pasted on the other side, when it is dried again. The under side is moistened at the edges and the surplus paper neatly rubbed off. When again dry the upper side is moistened all over and the paper over the top of each cracker is pierced with a punch or round stick so they may be charged with the necessary powder and clay. Some operators hold several sticks between their fingers at one time so as to be able to punch several holes simultaneously.

A wooden board about 1 inch wider than the bundle of crackers all around and $\frac{1}{4}$ inch thick with $\frac{1}{8}$ inch holes bored through it, corresponding exactly in position with the crackers in the bundle, is now laid on a smooth board, covered with finely powdered clay which is pressed into the holes with the hand, until it is firm enough not to fall out when the piece is lifted. The surplus is brushed off and it is placed over the bundle of crackers so that the clay-filled holes are exactly over the holes in the tubes. A slight blow is usually sufficient to cause the clay to fall into the crackers. Any not falling out is pushed out with a stick. The bundle is jarred slightly against the table to make the clay settle. A similar operation is now performed with a thicker board containing slightly larger holes, with the powder charge, after which the clay board is used once more as described above, filling the tubes completely.

The top layer of paper is now moistened so that it may be removed entirely and the clay, which has become slightly moistened as well, is gently pushed down with a suitable rammer. It is then dried in the sun. The bottom end is now carefully dipped into water, turned bottom up, and the paper removed from this side also, the clay is pushed down and pierced with an

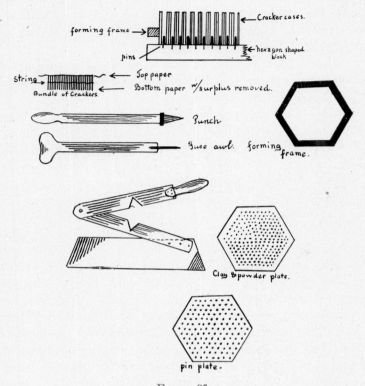

FIGURE 95

Firecracker Apparatus

awl for the purpose of inserting the match or fuse. However, this is not done until the crackers have again been dried in the sun. After the fuses are inserted, the ends of the crackers are pinched around them, about ⅛ inch from the end, by a crimper or by two blunt knives hinged together at one end and having a V-shaped notch cut out of the center of each blade so that when the two notches approach from opposite sides they pinch the cracker together and cause the fuse to be held in place. When they are now finally dried for the last time they are platted together to form the packs of commerce. The platting and wrapping of the cracker packages requires such dexterity that it is useless to try to describe it, as the necessary skill is only acquired after many years of practice.

The following formulas are in use for making the composition used in Chinese crackers and flash crackers:

CHINESE FIRECRACKERS

	Formula 1	*Formula 2*
Saltpeter	50	45
Sulfur	25	18
Charcoal	25	25
Potassium chlorate		8
Sand		4

FLASH CRACKERS

	Formula 1	*Formula 2*	*Formula 3*
Saltpeter	50	25	
Sulfur	30	25	30
Fine aluminum	20	25	40
Potassium chlorate	10	50	30

The fuse is a very important part of the Chinese cracker and is very difficult to make. Very tender and skilled fingers are required to produce this small yet most necessary adjunct. A thin strip of the finest Chinese tissue paper, about ¾ inch wide and 4 inches long, is laid on a smooth damp board; a little stream of powder is poured down its center from a hollow bamboo stick, and with the tips of soft-skinned fingers, which seem to have an attraction for the paper, it is placed against the lower right-hand corner. A rolling motion in the general direction of the upper left-hand corner causes the paper to roll up into a twine like fuse. The slightest touch of thin paste secures the end and prevents unrolling. When dry it is cut into the required lengths and is ready for use.[16]

[16] The information upon which the foregoing article was written has been supplied by Mr. Ip Lan Chuen, manager of the Kwong Man Loong Fireworks Co. of Hong Kong, China.

FIGURE 96

Girl Holding Finished Firecrackers

Simple Methods of Analysis for Fireworks Compositions

Probably every pyrotechnist, at one time or another, would like to know the composition of some product which he has seen for the first time and would like to be able to reproduce himself. During the many years that the author has worked with fireworks, there has been practically no single composition which he has not been able to duplicate by one means or another.

There is a decided disposition on the part of many pyrotechnists to consider their formulas as *valuable secrets* and usually those with the least technical knowledge place an added value on the little which they really know. If any fireworks maker has ever actually become wealthy through his secret formulas it would be interesting to learn his name.

Incidentally, all pyrotechnists of any standing, both in this country and in Europe or elsewhere, make relatively the same products and, with the widely disseminated knowledge of analytical chemistry, it is not difficult to analyze a composition.

With no pretense of giving a scientific scheme of chemical separation, the following suggestions have frequently been the means of determining accurately the composition of various mixings.

First, examine the physical appearance of the article or mixture in question. It is seldom pure white; but if it is, very little can be learned by looking at it. If it is black, it most likely contains charcoal, lampblack, antimony, meal powder or copper oxide or sulfide. If it is red, it probably contains red arsenic, red lead or red gum. If it is blue, look for copper sulfate or copper ammonium salts. If it is green, Paris green is almost sure to be present. If it is brown, asphaltum or various gums are most likely present. A yellow color indicates orpiment or sulfur.

A low-powered microscope is very useful in determining many of the ingredients of compounds. The substances are usually rather coarsely ground and individual particles can often be identified. This applies particularly to charcoal, aluminum, steel filings, shellac, red gum, copper salts, sulfur, etc.

To determine whether a compound contains chlorate of potash or other chlorates, it is only necessary to place a pinch of it on a slab and touch it with a glass rod dipped in sulfuric acid. It will burst into flames if chlorates are present.

Carbonates can be identified by adding a strong acid (nitric, sulfuric, etc.) to a solution or the substance under examination.

The presence of carbonates will be indicated by the copious evolution of carbon dioxide.

Sulfates may be detected by adding a few drops of a barium salt to some of the mixture which has been dissolved in water and filtered. A heavy white precipitate forms if sulfates are present.

To check for chlorides add a solution of silver nitrate to some of the dissolved and filtered compound. A white cheese-like precipitate indicates chlorides if the precipitate redissolves upon the addition of ammonia.

Nitrates are more difficult to identify because the presence of chlorates, which must always be suspected, interferes with the test for nitrates. If the test for chlorates, as directed in a previous paragraph, is negative, the following method may be used to detect nitrates.

Into a 6 inch test tube pour 2 ounces of a concentrated, filtered solution of the substance under examination. To this add, very carefully, little by little, an equal volume of strong sulfuric acid and allow to cool. Let the acid flow slowly down the side of the tube so that it goes to the bottom. Then, cautiously, add a concentrated solution of ferric sulfate so that the fluids do not mix. If nitrates are present the junction of the two layers shows at first a purple and afterwards a brown or red color. When the fluids are mixed, a clear brownish-purple liquid results.

If chlorates are present the following method must be used. Concentrate the solution by boiling; when it is cold add alcohol to assure removal of possible water-soluble gums. If cloudiness results, filter. Add an excess of sodium carbonate; evaporate to dryness and ignite gently to convert chlorates into chlorides. Now redissolve and test as directed in the previous paragraph.

Compounds containing ammoniacal salts give a brown precipitate with Nessler's solution. This test can be checked by adding a little dry slacked lime to some of the dry compound under examination, mixing thoroughly and moistening slightly.

The characteristic odor of ammonia is evolved if ammoniacal salts are present.

Flame tests, of course, are the simplest. In most cases, just igniting a little of the compound is all that is necessary. However, if this is impractical, dissolve a little; dip a platinum wire into the solution and hold it in the flame of an alcohol lamp. Strontium and calcium as well as lithium will color the flame red. Green indicates barium or boron; yellow, sodium; blue or light green, copper; pale violet, potassium, and white indicates antimony, arsenic, aluminum or magnesium.

After everything has been done to determine as completely as possible what is in the mixing, begin the quantitative analysis. For example, let us take a mixture which we have determined is red fire (a star, torch, etc.). As the average pyrotechnist is not an expert chemist with a well equipped laboratory, it is easier for him to work with larger quantities of the substance under investigation, than the chemist would ordinarily use. Twenty to thirty grams have been found to be a convenient amount, in most cases. Weigh this amount out accurately; place in a 6 ounce beaker and add 3 ounces of water. Warm slightly and filter, first by decanting the supernatant liquid, using a glass funnel and round filter papers about 6 inches in diameter. Add 2 ounces more of water to the residue and warm. Stir with a glass rod and wash it all onto the filter with a wash bottle, rinsing thoroughly on the filter and collecting all of the filtrate in the same receptacle. Label this *filtrate No. 1.*

Carefully dry the residue on the filter and weigh, placing a duplicate filter paper on the opposite scale pan to equalize the weight of the one on which the residue remains. The difference between this weight and the one you started with will represent the water-soluble portion of the mixing. Note this carefully and use it to check your final determinations.

Carefully remove as much of the residue as possible on the filter to a smaller beaker and treat with alcohol, warming slightly. After a half hour, pour it onto the same filter paper which has been returned to the funnel; wash with warm alcohol and label anything which might have dissolved *filtrate No. 2*. Dry the residue again and weigh as before; the loss gives the percentage of alcohol-soluble gum in the composition. This can be checked by evaporating filtrate No. 2 to dryness and weighing. The result should be the same as the second loss. If a residue remains, it can only be sulfur, the impurities of the gums or waxes insoluble in alcohol (which should be discernible under the microscope) and charcoal, lampblack, sawdust, etc. If paraffin or stearin are suspected, treat with gasoline as before and evaporate the filtrate for examination. If a black residue remains, boil it with hydrochloric acid. If there is no apparent loss, it is most likely charcoal or lampblack. If there is a loss, test the solution for copper, iron, lead, aluminum, etc. If sulfur is present, it may be removed by treating with carbon bisulfide and noting the loss as before. If the residue appears to be just an adulterant or impurity, it may be safely ignored. Simply note its weight.

Sometimes it is possible to detect iron by adding HCl and testing a few drops in a test tube with potassium sulfocyanide which gives a bright red color. Lead with potassium dichromate and a little acetic acid gives a beautiful yellow precipitate. If other substances are not present, ammonia will give a deep-blue coloration in presence of copper and a white gelatinous precipitate with aluminum. These reactions require checking, but they will sometimes put you on the track and are worth trying.

We must now return to *filtrate No. 1*. It can only consist of strontium nitrate, potassium chlorate or perchlorate, saltpeter and possibly water-soluble substances such as dextrin, cornstarch, gum arabic, etc. Evaporate the filtrate on a water bath until only about 1 ounce or less remains; cool and add alcohol. If the solution becomes cloudy, probably gum arabic or dextrin is present. If gum arabic is found, it may be removed and estimated by filtration, drying and weighing. If the solution remains clear, it may contain starch in some form and will present a viscous appearance. This may be determined by the addition of a small amount of a solution of potassium iodide and chlorine water. If starch is present, the solution will become indigo blue.

The remainder of filtrate No. 1 may now be evaporated to dryness on a water bath and weighed. The total weight added to the losses in the previous operations should equal the original amount used. This remainder will most likely consist of the salts mentioned at the beginning of the previous paragraph. The strontium may be separated as carbonate by adding ammonium carbonate until effervescence ceases. Collect the white precipitate on a filter, wash, dry and weigh. The weight will be sufficiently near to that of the nitrate to serve as a basis for calculating the final results.

To determine the amount of potassium salt present, evaporate the remainder of filtrate No. 1 to dryness, heating slightly until the ammonium nitrate is decomposed and dissipated. This operation is not entirely essential as simply adding up the total of the results obtained in the previous separations and deducting it from the weight of the original material will give the weight of the potash with sufficient accuracy to serve the purpose desired.

In analyzing green compositions, the same procedure may be followed using sulfuric acid instead of ammonium carbonate. By the addition of sulfuric acid to filtrate No. 1 the barium will be precipitated as the sulfate and may be collected and weighed as previously described.

In the analysis of blue compositions, the copper sulfide or oxide (the green parts) will be found in the first residue when the water-soluble ingredients are filtered off as described above. If a soluble copper salt has been used, it can be precipitated from filtrate No. 1 by adding excess sodium carbonate which will precipitate all the copper as a greenish-blue powder.

In mixtures containing saltpeter, sulfur and charcoal, in connection with steel filings, and/or aluminum, these can be quite accurately estimated by first washing out the saltpeter. Then the residue is dried in an evaporating dish and weighed. Treat the residue with hydrochloric acid, heating if necessary until it dissolves (filtrate No. 2).

Add a solution of caustic soda just to the point where no further precipitation occurs. This will precipitate both the iron and aluminum. The precipitated iron will not redissolve in an excess of the caustic soda but the aluminum will. It is therefore desirable to divide the filtrate No. 2 into two equal parts. Treat one half with caustic soda solution just to the point where precipitation ceases and treat the other half with an excess of the precipitant. If, upon weighing the precipitates obtained separately, the one to which the excess of caustic soda was added weighs less than the other half, this loss will represent the aluminum.

The sulfur and charcoal will be found in the second residue which has been treated with the hydrochloric acid. Wash thoroughly, dry and weigh. Treat with carbon bisulfide; wash, dry and weigh again. The remainder will represent the charcoal; the loss will be the sulfur. If magnesium is present, filtrate No. 2 is treated with sodium ammonium phosphate instead of caustic soda. Dry and ignite the precipitate and multiply its weight by 0.36 to get the magnesium content.

Pyrotechny in Warfare and Industry

The first use of fireworks in any form was most likely in military operations. Firecrackers in China, as reported by Marco Polo, may have been used in religious or other celebrations but, most likely, were first used by the Chinese armies, if only for the purpose of instilling fear in and demoralizing the enemy who had not heard them before, somewhat as the *screamer shells* were used in the beginning of World War II.

Greek fire, made of various materials, principally saltpeter, pitch and sulfur, was poured, burning, from the walls onto the assaulting troops. Sky rockets are recorded to have been used by the British, though not very successfully, in some of their wars. Mines and bombshells seem to have advanced both for pleasure and destruction during comparatively recent times.

In World War I, we probably saw, for the first time, the Very pistol and its modifications for firing signals by the use of stars of different colors. Airplane flares for signalling and parachute flares for illuminating areas to assist night attacks are also of recent use.

Smoke screens have probably been in use a long time but they have been developed to a wonderful degree of efficiency in recent years. An article of great importance and not used to any extent before this century is the *tracer bullet*. This is used to assist gunners in ascertaining the accuracy of their fire. It consists of small charges of composition (fire at night and smoke in the daytime), attached to projectiles and burning while the shell is in flight. By observing them the artilleryman can correct his sights if he is aiming too high or too low. Almost any of the formulas for fast-burning pumped stars are suitable for this purpose. The white candle star composition with arsenic would be good either for day or night use.

In industry, thermite and allied substances, consisting of chemicals which produce intense heat while burning, belong rather to the field of technical chemistry. However, many of the mixtures given in this book will melt iron while burning.

Tire Patches

The burning portion of self-vulcanizing tire patches is made by the use of a slow-burning composition of saltpeter and a carbonaceous material, such as sugar, moistened with gum water and packed into the tin receptacle.

Trick Matches

Trick matches are made by mixing silver fulminate with shellac solution and painting a minute portion upon the match stem ¾ inch below its tip.

Trick Cigars and Auto Back Fires

Trick cigars and auto back fires are also in this category. Detailed methods of manufacture should be given with discretion, as their use is not recommended.

GLOSSARY

Awl. Tool used for piercing lances.

Bengola. A ship light.

Black match. Bare or uncovered match.

Bounce. The report often used to end gerbes, saucissons, etc.

Bouquet. See *Flight.*

Candles. Roman candles. A paper tube from which stars are projected.

Cases. The containers, usually paper, in which pyrotechnical compounds are charged for producing the desired effects.

Chain. A garland of stars hanging from a parachute in the air.

Choke. The restricted opening at end of gerbes, etc.

Coal. Powdered charcoal.

Compo. Abbreviation of composition. Pyrotechnical mixtures.

Cone. The top of a skyrocket containing the heading.

Drift. A rammer.

Driver. A case of strong composition used for moving revolving devices.

Dross. The lava-like melted portion of tableau fires, etc., produced by the burning of their ingredients.

Feather edge. The margin of paper before the roll or sheet is trimmed.

Flares. See *Torches.*

Flight. A number of skyrockets (100 to 1,000) fired simultaneously; also shells fired in the same manner.

Former. The rod or tube upon which cases are rolled.

Fountain. Long gerbes containing colored stars, usually $1\frac{1}{2}$ inches in diameter and stuck into the ground for firing.

Fuse, blasting. Rubber-covered cotton tubing containing gun powder and used for slowly conveying fire to explosives.

Fuse, shell. A short paper tube of powder which burns while the shell is ascending, conveying fire to contents.

Gerbe. A pyrotechnical device usually $\frac{3}{4}$ inch in inside diameter from which a jet or spray of fire issues.

Grain (of paper). The longitudinal section of sheet as it comes from the rolls.

Gum. Dextrin, shellac, etc.

Gun. See *Mortar.*

Heading. The portion of a skyrocket which is released at the zenith of its flight.

Lances. Small paper tubes, ¼ to ⅜ inch in diameter, charged with composition and used for producing outline pictures in fire.

Lancework. Letters, designs, etc., produced in fire by the use of lances.

Leader. The long section of match from which pyrotechnical pieces are lighted.

Match. Cotton wicking impregnated with gun powder, used for conveying fire to a piece or parts of pyrotechnical devices.

Match, black. See *Black match.*

Match pipes. Paper tubes into which match is inserted. The tube protects the match and makes it burn much faster.

Meal. Finely powdered gun powder.

Mines. Paper mortars charged with stars, etc.

Mortar. A metal or wooden tube from which bombshells are fired.

Nipple. The device which forms the choke of gerbes, drivers, etc.

Nosing. Two or three turns of paper projecting from the ends of gerbes, etc., used for securing the match when matching pieces.

Port fires. Paper tubes, ⅜ inch in diameter, charged with a composition which, while burning, is used for lighting other pyrotechnical devices.

Priming. A sludge made of gun powder and gum moistened with water.

Quick match. Match covered with paper piping.

Rammer. Wood or metal rod used in ramming cases.

Ramming. Charging cases with fireworks compounds.

Revolving pieces. Pyrotechnical devices consisting of movable parts.

Rolling. The act of producing fireworks cases from paper and paste.

Saltpeter. Potassium nitrate (niter).

Serpents. Nigger chasers; Pharaoh's serpents.

Set pieces. Devices consisting of lancework, gerbes, etc.

Shells. Pyrotechnical devices which explode in the air and are fired from mortars.

Spindle. A tapering metal rod which provides the hollow center of rockets; the bearing upon which revolving pieces operate.

Stars. Small masses of pyrotechnical compounds which are discharged from shells, candles, etc., and burn while in the air.

Striking fire. The act of accidentally causing an explosion while working with pyrotechnical compositions.

Temple piece. Pyrotechnical ensemble consisting of batteries, mines, bengolas, etc.

Torches. Paper tubes, $\frac{1}{2}$ to 1 inch in diameter, charged with pyrotechnical compounds, usually colored and used in parades, etc.

List of Pyrotechnical Books

Paul Tessier, *Chimie pyrotechnique* — 1883
Domenico, Antoni, *Trattato de Pirotechnia Civile* — 1893
J. R. J. Jocelyn, *Royal Fireworks* — 1906
H. B. Faber, *Military Pyrotechnics* — 1919
A. St. H. Brock, *Pyrotechnics* — 1922
G. W. Weingart, *Manual of Fireworks* — 1938
Tenney L. Davis, *Chemistry of Powder and Explosives* — 1940

INDEX

A

B